Louis Coatalen

Louis Coatalen
Engineering Impresario of Humber, Sunbeam, Talbot and Darracq

Oliver Heal

UNICORN

Contents

	Preface	7
	The Coatalen Family Tree	10
	Introduction *A Man in a Hurry*	13
One	Pirates, Parents and Preparation	17
Two	Coventry Calls *1901–08*	28
Three	Wolverhampton Wonders *1909–14*	50
Four	The First World War *1914–18*	106
Five	A New Beginning? *1919–23*	137
Six	Blowers and the Seeds of Destruction *1924–29*	175
Seven	Collapse and Recovery *1930–39*	205
Eight	The 1940s and After *1940–62*	237
	Appendix One – Personal Competition Achievements	256
	Appendix Two – Patents	264
	Sources	273
	Index	275

Preface

Louis Coatalen, as Chief Engineer of the Sunbeam works, was undoubtedly a hero for my father, Anthony S. Heal, who had been a schoolboy when Sunbeam racing cars were triumphing in the UK and abroad. He had followed closely their adventures by studying *The Autocar* weekly when he should have been fielding on the cricket pitch or doing his homework. As an adult my father formed a collection of some of the key Sunbeam racing cars and spent many years researching and recording their history. This fascination finally resulted in him writing the definitive book on the subject, *Sunbeam Racing Cars 1910–1930*. I was therefore brought up in an environment imbued with Sunbeam racing cars, their histories and a respect for the man responsible for their creation.

Many years later, when I told my father that I was going to marry Annik, one of Louis Coatalen's granddaughters, he was 'over the moon'. I do not think he could have been more delighted if I had married a member of the royal family!

In contrast, it came as something of a shock to discover that my father-in-law, Hervé Coatalen, in no way shared this admiring vision of his own father. He had not had a happy relationship with him. Hervé's parents had separated when he was seven years old, so his childhood memories of his father were mainly of being taken around factories during school holidays. Later he had worked for him but found him an erratic tyrant. He was certainly not keen that any eulogistic biography of Louis should be published.

However, encouraged by several of Louis' granddaughters, I eventually set out to produce a record of Louis Coatalen's life which I trust is sufficiently well-balanced to portray the man accurately, so that both the Coatalen and Heal families would recognise the personality at its heart. Over many years I have collected as much information about different aspects of his career and his life as I could find, returning wherever possible to primary sources to produce a reliable record of an extraordinary life.

O.S.H.

Acknowledgements

My thanks go to the Coatalen girls for all their help and their patience in waiting for this finally to be published. Particular thanks to Carole de Chabot, who unearthed much material and provided contacts, as well as being consistently encouraging and supportive over the years. I am sorry it took so long!

Much information came from the Heal Sunbeam archive, as well as from Coatalen family papers. The archives held on the Sunbeam Talbot Darracq Register Forum website (and the physical archives) have been most useful in addition to the information provided by individual members such as: Bill Barrott, Jim Catnach, Bruce Dowell, Pat Durnford, James Fack, Sebastien Faurès Fustel de Coulanges, Ian Goldingham, Stephen Lally, Karl Ludvigsen, Alan Richens, Ian Walker, Andy Watt and Ben Yates. Heartfelt thanks to all.

The late John C. Tarring was a valuable source of Humber information. Huib de Vries in Holland provided the information on *Tirrena*. In France, in addition to Carole de Chabot, Annie Mell, Michel Guegen, Françoise et Jean-Michel Gloux, M. Hetzlen, Emmanuel Piat and Daniel Richard were all most helpful for which I am very grateful. My apologies to anyone I have forgotten to include. I am very thankful for the efforts of Jenny Symon, Stephen Lally and Ben Yates who read the draft and picked up my errors of spelling and substance and suggested valuable improvements. Any faults that remain are my responsibility entirely.

Peter Card of the Society of Automotive Historians in Britain and the Michael Sedgwick Memorial Trust has been a fund of practical advice about publishing and I am most grateful to the Trustees of the MSMT for their generous contribution to the cost of producing the book. Ian Strathcarron, Lucy Duckworth, Felicity Price-Smith and the team at Unicorn have done an excellent job of making it into a really attractive publication.

Finally, many, many thanks to Annik for putting up with me spending hours in front of the computer or researching at length some bizarre twist of the story and thus not contributing much to family life. It would never have been completed without your valuable support.

MICHAEL SEDGWICK
MEMORIAL TRUST

Louis Coatalen is published with the financial assistance of the Michael Sedgwick Memorial Trust. The M.S.M.T. was founded in memory of the motoring historian and author Michael C. Sedgwick (1926–1983) to encourage the publication of new motoring research and the recording of Road Transport History. Support by the Trust does not imply any involvement in the editorial process, which remains the responsibility of the editor and publisher. The Trust is a Registered Charity, No 290841, and a full list of the Trustees and an overview of the functions of the M.S.M.T. can be found at: www.michaelsedgwicktrust.co.uk.

The author is also most grateful to the following subscribers who have kindly contributed to the cost of publication:

The Sunbeam Talbot Darracq Register
T.J. Cardy
Andrew Crisford
Peter & Sally Dodds
Stephen Enthoven
Paul Grist
Felix Heal
Simon Hinson

The Hunter Family
Robert Mansfield
Craig McWilliam
Tim O'Brien
Ian S. Polson
Alan Richens
Michael Tyler
Andrew Watt

Illustrations

Most of the photographs are from the Coatalen and Heal family collections. Thanks are due to Adel Hanna for re-photographing a considerable number from the Olive Coatalen scrapbooks. Other illustrations come from: *The Autocar* (51), Automobile Club de France (Emmanuel Piat) (245, 252), Bibliotheque Nationale de France (139, 143,158), Alec Brew (110), Bruce Dowell (53, 105, 145), Pat Durnford (150), Mary Evans Picture Library (174), Espacenet.com (116, 117, 201, 229, 230, 247 and Appendix 2 is extracted from the European Patent Office website), Diana Gibbs (172), Françoise et Jean-Michel Gloux (22), Grace's Guide (30), Adel Hanna (115, 244), H.M. Hobson (96), Stephen Lally (162), Bruno Lovera (179 top l.) *The Motor* (63), National Motor Museum (60, 192, 208), Richard Roberts Archive (242, 246), STD Register Photographic Archive (Carol Allen) (85, 111, 120, 124, 125, 126, 206), John C. Tarring (32, 38), Huib de Vries (235, 236), Ben Yates (71 top), Keith Yeo (113, 114). All of whom are warmly thanked. Should any copyright material have been used inadvertently please accept my apologies.

The Coatalen Family Tree

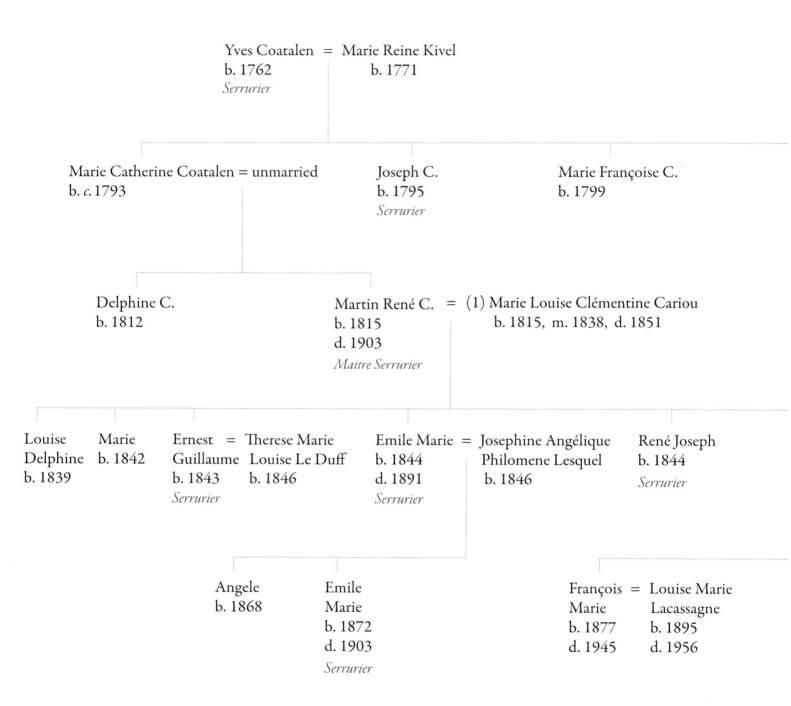

Jean Pierre C.
b. 1801

Yves François C.
b. 1812

= (2) Marguerite Clémence Le Saour
b. 1823, m. 1850s

François Marie = Louise Marie
b. 1850, Angélique Le Bris
m. 1876 b. 1850, d. 1893
d. 1905
Charron

Marie
b. 1860

Leontine Emilie
b. 1863

Louis Hervé
b. 1879
d. 1962

= (1) Annie Davis
b. 1883
m. 1902

= (2) Olive Mary Bath
b. 1891
m. 1909
d. 1969

= (3) Iris van Raalte
(née Graham)
b. 1891
m. 1923
d. 1975

= (4) Ellen Amy
Bridson
b. 1896,
m. 1935
d. 1973

Louis Hervé
b. 1905
d. 1971

Hervé Louis
b. 1913
d. 1999

Jean Louis
b. 1916
d. 1976

Marjolie
b. 1924
d. 2013

Louis Coatalen 1879 – 1962

Introduction

A Man in a Hurry

The life story of Louis Coatalen is remarkable because he originated from a modest background in Brittany, the westernmost part of France, and rose to be one of the leading motor manufacturers of his time in Great Britain. He was undoubtedly a man of great drive and determination, which he combined with Gallic charm and wit. He was a 'Man in a Hurry', not just because of his fascination with speed and his determination to win, but also because, as a young man before World War I he speedily brought considerable success to two different manufacturers, thus firmly establishing his reputation as an important engineer. Later he led a life that involved not only fast cars that won Grand Prix races and set up World Land Speed Records, but one which also included stories of industrial espionage, beautiful women, drugs and expensive yachts.

Having reached his peak, Coatalen returned to France and, after some years, went into a sad decline before making a comeback in later life. The aim of this book is to pull together the varied elements of his life that have become scattered on either side of the Channel in order to present as complete a picture of the man as is possible. His contribution as an engineer has been very largely covered in books such as *Motoring Entente*, *Sunbeam Racing Cars*, *Sunbeam Aero-Engines* and *The Humber Story*, so it is unnecessary to repeat here the technical details of the products for which he was responsible. Instead, I have tried to fill in some of the lesser-known aspects of his career and provide a fuller portrait of this fascinating individual.

At the end of World War I, when he was thirty-nine years old, Coatalen was described as 'a dark, clean-shaven young man of a fresh complexion, with a quietly confident manner and humorous brown eyes of more than common intelligent expression.... Louis Coatalen is blessed with more than average good physique,

Introduction

a fact which is further confirmed when one observes the manner in which his head is set on his shoulders and the broad base of the skull.'[1]

There are two Coatalen family stories that illustrate well what a competitive character he was. The first concerns Louis Coatalen as a teenager. At the end of September every year the Tréminou fun fair, which has a tradition that stretches back to long before the French Revolution, is held in Pont l'Abbé, the town where Coatalen spent part of his boyhood. By the end of the nineteenth century people came from far and wide to join the celebrations, which included horse and bicycle races that awarded attractive cash prizes to the winners. Louis took part in one of the bike races around the main square and, as he battled for the lead with another lad from the town, he said: *'Laisse-moi gagner et je te donne la moitié'* (let me win and I'll give you half). It seems that he did win and he and young Larnicol split the proceeds between them.

The other story concerns him as an old man but indicates that even then he retained his competitive edge. It is well known that the French greet each other with a kiss on both cheeks, but in Brittany for members of the family the dose is doubled. His granddaughter, Annik, remembers how, when she was a little girl, Louis's approach to greeting granddaughters was to peck them on the cheek four times as quickly as he possibly could and then cry, *'J'ai gagné!'* (I've won!).

Further indications of the character of the man can be gathered from descriptions left by people who knew him. For example, W.O. Bentley, a competitor and contemporary, wrote in his memoirs about Louis Coatalen that he was 'vivid and exotic'. Over the years they came into frequent contact, but it all began at the 1914 Tourist Trophy race where they stayed in the same hotel. According to Bentley, this was 'quite a social affair' that provided plenty of time for 'the entrants, drivers, manufacturers and mechanics to get to know one another'. Subsequently, during World War I, Bentley was put in charge of informing aero-engine manufacturers about the advantages of aluminium pistons. He called on Louis Coatalen at Sunbeam in Wolverhampton whom he describes as squat, jovial with a puckish grin, highly amusing and a tremendous raconteur. 'One of Coatalen's most endearing qualities was his absence of illusions about himself.... I think Coatalen would accept the fact that he was not a great designer, but he had many characteristics of a first-rate designer, including intelligence and the ability to learn lessons of others.'[2]

Behind the charm was a great deal of determination. Pigheadedness is a characteristic often attributed to the inhabitants of Brittany, and Coatalen's friend, the famous French motoring journalist, Charles Faroux, wrote that his Breton roots were often cited fondly by his friends as an explanation of his perseverance and obstinacy that drove him throughout his career. Even if one might not call him pigheaded, having had an idea based on sound thinking he would defend it and maintain it with determination.

---

1. *The New Zealand Motor & Cycle Journal,* 25.10.1918. 'An Aircraft Engineer – Louis Coatalen's Meteoric Career'. Thanks to Ian Walker for drawing my attention to this article and to Ian Goldingham (NZ STD Reg.) for providing a copy.
2. W.O. Bentley, *The Cars in my Life*, Hutchinson, 1961, pp. 69-73 and W.O. Bentley, *W.O. An Autobiography*, Hutchinson, 1958, p. 76.

Faroux explains that Coatalen had a tremendous facility for assessing the capabilities of workshops and factories. He was known amongst his friends as *'le dénicheur de loups'* (the mess-up spotter). 'When Coatalen visits a factory he invariably falls upon the one thing he is not meant to see; he notices immediately the weakness of organisation, the machining fault, the hold-up in supplies to the machines. He spots it instantly. But his comment, even if it is ironic, is always benevolent.'[3] This fastidiousness is confirmed by an ex-Sunbeam employee who recorded that Coatalen was a hard taskmaster and kept everyone on their toes. 'He was very precise in all he did. When walking around the machine shop he would look into the operator's cupboards and if he found all equipment and tools in orderly array he would have a half-crown put in the man's pay tin. If he saw tools in disarray he would recommend a severe reprimand.'[4] He had a reputation for being very caring about his employees.

S.C.H. Davis, who was for many years Sports Editor of *The Autocar* but who had known Coatalen since 1910 when Davis joined *The Automobile Engineer* as an illustrator, described how this job brought him into contact with the designers of the world's best cars whom he found 'very interesting indeed and very human'. He wrote, 'I took a special liking to Laurence Pomeroy of Vauxhall and Louis Coatalen of Sunbeam, both of whom tempered high technical knowledge with humour and, in a way, extravagance.'[5] 'In an argument Louis would get over any strong opposition by talking fast in a mixture of French and English for long enough to think out the right repartee, to which there was no counter.' Davis recorded how he enjoyed hearing 'Louis having a discussion with Bugatti about design, that being like a fight with rapiers, quick thrust following quick thrust, each enjoying the game',[6] and Davis also described to Griff Borgeson how Louis could extract information from designers such as Ernest Henry:

> Henry would pinch anybody's ideas, and he was damned clever. But he was no match for Coatalen. You kept feeling, 'I know what Louis is trying to do. If Henry doesn't watch it, in one of these talks with Louis he'll let it all out.' Then before you knew what was happening, Louis would show him a sketch he claimed to have done three years ago, showing how he happened to have been thinking along the same lines. And suddenly the two of them would be discussing it and Louis would go away with every single thing he wanted.[7]

From Henry Segrave we get the impression that Coatalen was direct and blunt in his dealings with people. On leaving the armed forces, Segrave wanted to get into motor racing and decided that driving for Sunbeam should be his target. He

---

3. Charles Faroux, *Journal de la S.I.A.*, July 1953.
4. Bill Fowler, *Memories of a Sunbeam apprentice*, unpublished, STD Register archive.
5. S.C.H. Davis, *My Lifetime in Motorsport*, Herridge & Sons, 2007, p. 35.
6. S.C.H. Davis, 'Pioneer Memories', *The Autocar*, 8 September 1979. p. 21.
7. Griffith Borgeson, *The Classic Twin-Cam Engine*, Dalton Watson, 1981, p. 106.

Introduction

eventually went on to be one of Sunbeam's most successful drivers. He admired Louis Coatalen, 'whom I do not hesitate to describe as one of the outstanding figures in British automobilism. Few designers of cars unite technical ability with sound business sense and sufficient dynamic energy to bring them into a high position of control in the motoring industry. Mr Coatalen is certainly one of them.' However, getting into the team was not easy. Segrave pestered Coatalen whenever he could to allow him to drive a Sunbeam. 'I was woefully lacking in road experience, and he made no bones about telling me that he did not believe I should be any good in a road race. I think it was even possible that it was solely as a means to get rid of me that Mr. Coatalen finally agreed to let me have a car. He made no secret of the fact that he thought I should be a failure.'[8] Having been given the opportunity Segrave was of course able to prove that he could be very successful.

We also get some insight into Coatalen's character from *The History and Development of the Sunbeam Car 1899–1924* by Massac Buist, who wrote that Coatalen's

> mind has been cast always on big lines. In him there is nothing of the small man whose sole aim is to impress a board of Directors with the idea of his own talent on the principle of consistently barring others from a protected territory. On the contrary, wherever Mr Coatalen sees a special talent, forthwith he endeavours to engage it because in his opinion it is impossible to have any department too well manned. He realises that an intelligent directorate will appreciate always that a big man's job is to get the best men procurable for every phase of its company's work, and that an organiser proves his ability in measure as he surrounds himself with strong and big men, as distinct from insignificant sycophants.[9]

Finally, an impression of the whirlwind that Coatalen was at his peak comes from an extract from an article published after the announcement that he was to be appointed a director of Sunbeam at the end of 1911. The journalist wrote of Louis Coatalen:

> He is a sort of Admirable Crichton. He is the chief engineer of the company; he is responsible for the design of the cars, also he is what the Germans would call the over-director of all the producing departments of the factory; he buys the materials, selects all the heads of departments in the works, arranges the output and, quite apart from all that, finds any amount of time for private touring, for golf, for local competitions, and for racing at Brooklands, not omitting the astonishing twelve hour record with a six cylinder Sunbeam planned and carried out by him. Yet he never seems to be in a hurry and is withal quite one of the most humorous conversationalists of the many merry men who compose the motoring fraternity.[10]

---

8. Sir Henry Segrave, *The Lure of Speed,* Hutchinson & Co., 1932, p. 87.
9. Massac Buist, *The History and Development of the Sunbeam Car 1899–1924*, The Sunbeam Motor Car Company, Wolverhampton, 1925, p. 24.
10. Unidentified paper cutting in Olive Coatalen album. *The Admirable Crichton* was a stage play written by J.M. Barrie in 1902 in which Crichton emerges from the role of butler to be the leader of the group when shipwrecked on a desert island.

One

Pirates, Parents and Preparation

The name Coatalen in the Breton language means 'wood by the lake' and underlines the deep roots of Louis Coatalen's family in Brittany. His parents were born in Pont l'Abbé in the Pays Bigouden, South Finistère but it is possible that the family originated from further north. Could it be that they were descended from the famous fifteenth-century privateers and merchants, the Coatanlems? Some might be tempted to see parallels with the sometimes rather piratical nature of Coatalen's dealings in the motor industry but the most important similarities lie in the manner in which the Coatalen family refused and still refuses to regard the English Channel as a barrier.

Pirates

The fifteenth century was perhaps the most brilliant period in the history of Brittany, when it was still largely independent of France and ruled by Duke François II (1435–88).[1] It was only at the end of that century that his daughter, Duchess Anne of Brittany (1477–1514), married King Charles VIII of France and after his death she re-married his successor, King Louis XII. Even so, it was not until 1532, long after she died, that Brittany was officially ceded to France.

So although Brittany was closely linked to France, it was still basically a separate country when Yann Coatanlem was born in Saint-Pol-de-Léon about 1455. He became a skilled and fearless navigator operating from the north coast ports of Morlaix, Roscoff and Bréhat Island with his own fleet of five boats. From there he kept an eye on traffic in the Channel and regularly raided foreign boats, returning home with his booty. In France to this day, Yann Coatanlem (or Jean de Coetanlem, as he

---

1. François II became Duke in 1459.

Chapter One

is called in French) is reputed to have attacked Bristol in 1484. Although outnumbered five to one, he is said to have captured a number of English ships that had sailed out specifically to do battle with him. He is then supposed to have sailed right into Bristol harbour, burning and pillaging the town and capturing prominent citizens for ransom.

However, the distinguished historian C.S.L. Davies, who has researched the subject in depth, suggests that the story has become exaggerated over time. That Coatanlem was a redoubtable sailor is not in doubt, nor is the fact that at the beginning of 1484 'a vigorous naval war was raging between England and Brittany'. It seems that Coatanlem's activities at that time were even encouraged by the Admiral of France who provided weapons with the result that he managed to capture considerable booty and three large ships from the English. However, no evidence has been found of an actual invasion of Bristol. By the middle of 1484 the political scene had moved on. A truce had been arranged between Brittany and England whilst relationships between France and England were becoming increasingly hostile.[2] Yann Coatanlem became something of an embarrassment, so it appears that, for a while at least, he was despatched to Portugal to assist the young King João II in the protection of his merchant fleet.

Yann had a nephew, Nicolas Coatanlem, with whom he was closely associated and who had taken part in some of his raiding activities. Nicolas was not much younger than Yann but developed to be a respected merchant and part of the minor nobility of Brittany, speaking Breton, French, Latin, English and Spanish. He built up his own trading fleet that exported salt from Guérande and sailcloth from Morlaix to Spain, returning with Spanish and Bordeaux wines or merino wool from Bilbao for the Breton weavers. Spanish fruit and wine were shipped to England and northern Europe, whilst Welsh coal was brought back to fuel Breton forges. However, in that period one could not be certain of being able to trade peacefully all the time and on a number of occasions Nicolas took part in the defence of his country to assure the independence of Brittany. For example, in 1487 when the French army laid siege to Nantes, Nicolas broke through the lines with some of his boats loaded with provisions for the inhabitants of the town. His importance as a merchant is demonstrated by the fact that in that same year King Henry VII of England issued him with a guarantee of safe passage. In 1495 Duchess Anne and Charles VIII put him in charge of building one of the largest and most modern ships of the epoch. The *Marie Cordelière* was of 600-tonnage capacity, capable of transporting around 700 men, well-armed and equipped with sixteen big canons. She was launched in 1498 in the Morlaix estuary and finally went down in 1512 in a battle off Brest, taking the English vessel *Regent* with her. Not much is known of Nicolas Coatanlem's later life but he died in 1519 at his Manor of Penanru on the Morlaix estuary.[3] His Will underlines his wealth and the extent of his international relationships. His principal customers were

---

2. C.S.L. Davies, 'The Alleged Sack of Bristol: International Ramifications of Breton Privateering, 1484–5', *Historical Research*, Vol. 67, October 1994.
3. Information from: Roparz Omnès, *Les Koatanlem, Marchands, Pirates et Patriotes Bretonnes, 1455–1519*, Editions Sked, 2000.

English buyers of linen cloth woven mostly at Locronan and he left gifts to religious houses in Brittany, Normandy, Paris, Spain and Walsingham in Norfolk, England.

Direct Family

There is no positive evidence to link these illustrious Coatanlems to the more modest forebears of Louis Coatalen some 300 years later, other than a similarity of name. However, it is a fairly unusual name, even in Brittany. As will be seen, in the nineteenth century the family were not swashbuckling mariners but skilled craftsmen in metal with their feet firmly on dry land, while being equally proud of their Breton heritage.

Louis Coatalen's ancestors have been traced back in a continuous line to the end of the eighteenth century. Although the developments of the twentieth century meant that Louis was involved in technologies that had not previously existed, his career as an engineer was in fact continuing the family tradition of skilled metalworking for a fifth generation.

The first direct ancestor that we know of was Yves Coatalen, Louis' great-great-grandfather, who was born in 1762 in the village of Plouhinec, not far from Pont Croix. Yves Coatalen lived and worked in Pont Croix, which was an important historic market town from the Middle Ages onwards with its port on the Goyen river estuary and an impressive church whose architecture inspired many others. Its large cattle market and narrow streets packed with shops served as the crucial trading centre of the windswept Cap Sizun area, the most westerly point of South Finistère. Nowadays its function as a port has been taken over by Audierne, nearer the mouth of the river, but in the eighteenth century it was an important place in the area. Yves Coatalen was the town's *serrurier* – a skilled wrought-iron worker, blacksmith, locksmith, etc., who also knew how to read and write. He had at least three sons and one of them, who was also a *serrurier*, appears to have moved about 20 miles further south-east to set up his forge in the town of Pont L'Abbé. Yves also had a daughter Catherine (born *c.* 1793) who worked as a grocer in Pont Croix. Although she was not married she had a daughter Delphine (b. 1812) and a son, both of whom were given their mother's surname as no father's name is recorded. The son, Martin René Coatalen (who was to be Louis' grandfather) was born in 1815. We do not know if he was trained by his grandfather Yves or by one of his uncles, but he too became a metalworker. From the 1836 census, when he was twenty years old, it emerges that he was then working as a *serrurier* and apparently running the forge in Pont l'Abbé as his uncle had died the previous year leaving behind a widow, Marguerite, and five young children. A couple of years later Martin married Marie Louise Clémentine Cariou and set up his own household in the Place du Marc'hallac'h (now Place Gambetta), Pont L'Abbé. Together they had two daughters and four sons between 1839 and 1850 but in 1851 Marie Louise died, leaving Martin with six young children. He soon got married again to Marguerite Clémence Le Sauour and later they had two further daughters.

Chapter One

During the nineteenth century the town of Pont l'Abbé was growing quickly whilst Pont Croix became a backwater. The decision to move had proved to be very wise. Between 1800 and 1870 the population of the town doubled from 10,000 to 20,000. Strategically placed to serve the ports of Guilvinec and Loctudy that were rapidly expanding, it was a commercial centre that exported wheat, potatoes, chemical products and salted fish. The railway came as far as Pont l'Abbé in 1884, opening up yet further market possibilities. Throughout this period Martin Coatalen exercised his metalworking skills. Family lore recounts that when the Augustine convent was installed in the town in 1860, the elegant wrought-iron railings and gates at its entrance were the work of the Coatalen forge. Martin lived to be eighty-eight years old and by the time he died in 1903 he would have been aware that his grandson Louis was already making a name for himself in England in the new world of the automobile.

Martin's three eldest sons, Ernest, Emile and René, followed in father's footsteps and became *serruriers*. The youngest son, François Marie (Louis Coatalen's father) born in 1850, whose mother had died when he was hardly a year old, learnt a similar but slightly different trade. He grew up to be a tall (1.90 m) dark-haired man with dark eyes and became a *charron* – which translates as a wheelwright, but the

Above: Map of Brittany.

Left: Mme Coatalen (née Louise Le Bris, b. 1853 d. 1893) mother of Louis Coatalen, who ran Hôtel de France in Concarneau from 1881 to 1893.

wheelwright's forge would have been expected to handle all sorts of repairs to carts and to shoe the horses' hooves. Perhaps because there were already so many Coatalen metalworkers operating in Pont l'Abbé he decided to move yet further along the coast to the port of Concarneau soon after his marriage. His bride, Louise Marie Angélique Le Bris, who gave her profession as *tailleuse* (tailor/seamstress), was twenty-four years old (b. 21 June 1853) when she married. Her father Jean François Marie Le Bris had died on 10 April 1869 but her mother Jeanne Marie Perrine Chéolade was still alive and they lived not far from Martin Coatalen in the rue des Carmes in Pont l'Abbé.

The first son of François and Louise was named, predictably, François Marie Coatalen, and he was born in Concarneau in 1877. Louis Hervé Coatalen, the subject of this biography and their second son, was born there a couple of years later on 11 September 1879. According to local historian Michel Guegen, the young couple acquired some nearby land from the convent and built the Hôtel de France at 11 avenue de Quimper, Concarneau in 1881. On the ground floor was an office, the kitchen and a café, whilst upstairs were eight guest rooms (four per floor). The wheelwright's forge was in the yard at the rear of the premises. The hotel's prominent position was made even better in 1883 when the railway came to Concarneau, as then all travellers arriving by train had to pass the hotel on their way into town and the road was renamed *Avenue de la Gare*. A number of painters made it their summer quarters, with the result that the walls were hung with works by artists such as Frederick Richardson, Emil Hirschfeld, Emile Schuffenecker and Charles Fromuth. Sadly, by then François Marie (Louis' father) had been admitted to the St Athanase mental home in Quimper in 1882, aged thirty-two, as his alcohol addiction had become so bad that he suffered hallucinations and delusions. He was to remain there, a somewhat tortured soul, until he died in 1905 aged fifty-five.

Louis Hervé Coatalen as a young boy.

His wife Louise continued running the hotel on her own, as is confirmed by a surviving letter-heading on which is printed *Hôtel de France, Avenue de la Gare*, below which appears in even larger letters, *Tenu par Mme COATALEN*. It seems likely that Louise had financed the project in the first place as she had received, as did her three sisters, the sum of Fr. 12,325 from her mother in 1880 in connection with the settlement of the estate of Mr Luc Roblin.[4] She also paid for her husband's upkeep at the hospital.

Louise was obviously a very caring person who communicated her warm-heartedness to her guests as well as to her family. Charles Fromuth (1858–1937), the American-born painter whose pastel drawings of Concarneau harbour have since become highly regarded, recorded in his diary that her 'interest and

---

4. Luc Roblin of Quimper was a witness at the marriage of François Marie and Louise Marie.

Chapter One

Top: The Hôtel de France was the preferred base for artists visiting Concarneau to paint and draw. This photograph taken in 1911 shows artists and staff on the pavement outside. Charles Fromuth, who taught the young Coatalen boys drawing and English sits on the kerb smoking a pipe.
Bottom left: The Hôtel de France as it must have looked when owned by the Coatalen family.
Bottom right: This 1927 postcard gives the view looking up the Avenue de la Gare towards the station with the hotel about halfway up on the left.

Hôtel de France letter-heading

sympathy towards me' when he first came to Brittany in 1890, 'was the fundamental cause of a home attachment to Concarneau' and the reason why he remained there until his death. In return for her hospitality he gave 'lessons in drawing and English to her two boys' which may well explain why they both later happily moved to England.[5] At the time Louis was a pupil at the local École Primaire Supérieure where he played the triangle in the orchestra.[6] Life at the hotel must have been quite lively as, in a letter to Fromuth, Louise excused herself for not writing a longer letter as wedding guests had been dancing through the night until nine-o-clock that morning.

Louise, however, died on 31 October 1893 when she was only forty years old and her sons were aged sixteen (François) and fourteen (Louis) respectively. Fromuth wrote in February 1894 to another American artist friend, Augustus Smith Daggy, about the loss of his 'beloved friend Madame Coatalen' and explained that he was looking after her grave 'for the boys are no longer here. I assure you it is continually covered with flowers and wreathes in which I display my taste and affections that I have never and shall never repeat. Surely I can say no one ever kindled in me so lofty an affection and once established produced such good results to my progress.'[7]

Louis, who had effectively lost his father

Louise Coatalen on her deathbed, 31 October 1893.

---

5. Charles Henry Fromuth, diary entry for July 1906 after Louis Coatalen had been to visit him. He noted that he 'has become a celebrity in automobile construction in Coventry England and visited me with three personages in his affair'. My thanks to Françoise and Jean-Michel Gloux of Concarneau for allowing me access to Fromuth's diaries.
6. Letter from Michel Guegen, 22 October 2009.
7. My thanks to Andy Watt for providing a transcript of the original letter in his possession dated 6 February 1894.

Chapter One

when he was just three and had now lost his mother when he was fourteen, thus became an orphan. His mother's brother-in-law, Hervé Julian Le Moigne (married to Louise's sister Marie le Bris), officially became legal guardian for the young Louis Coatalen on 13 August 1894 but evidently the two brothers had already moved to Pont l'Abbé at the end of 1893 to be brought up by their aunt and uncle. The hotel was sub-let and eventually sold in 1899 for Fr. 24,100. The Le Moignes had three children of their own: Jeanne (b. 1878), Marie (b. 1882), and Julien (b. 1886), with whom the Coatalen boys grew up. We know that Louis remained in contact with these cousins in later life as, during World War I, he received photographs of Jeanne working as a nurse in Pont l'Abbé and Julien in uniform at the wheel of a mobile searchlight unit.

It was presumably Hervé Le Moigne who made the decision to send Louis Coatalen away to complete his education at the École des Ouvriers et Contremaîtres at Cluny after he had attended the Lycée at Brest and helped him obtain a place there.

Gadzarts

In later life Coatalen made much of his training as a *Gadzarts*, as the engineers from the École d'Arts et Métiers were known, and he seems to have been proud to be one of the alumni of this establishment. In the booklet produced by Sunbeam entitled *A Souvenir of Sunbeam Service 1899–1919*, it is recorded that the three-year course kept him busy 'from early dawn to dewy eve, seven days a week with a half day off on Sunday, plenty of preparation work to do, and a fortnight's holiday in the year'.[8] It certainly seems to have been a tough and thorough course that imposed discipline and demanded dedication from the students in their military style uniforms.

There were four *École d'Arts et Métiers* (School for Skills and Trades) in France. The original schools had been founded by Napoleon but the one at Cluny, which Coatalen attended, had opened in 1891 so was relatively new.[9] In fact, when he went there it

M. Hervé Le Moigne, who was Louis' guardian, and his wife Marie, with their grandson, Hervé, born in December 1915.

---

8. *A Souvenir of Sunbeam Service 1899–1919*, Sunbeam Motor Car Co. Ltd., 1919, p. 13.
In an article published in the *NZ Motor & Cycle Journal*, 25 October 1918, the hours of work at Cluny were given as 5.00 am to 9.00 pm seven days a week.
9. The first school was established at Compiègne in 1803 but transferred to Châlons-sur-Marne in 1806, the second at Angers in 1811, and the third at Aix-en-Provence in 1843. Under the 3rd Republic Cluny (1891), Lille (1900), and Paris (1912), were added. These secondary schools were transformed into engineering institutes in 1907 and promoted to university level in 1945.

Julien Le Moigne at the wheel of a mobile searchlight vehicle, 1916.

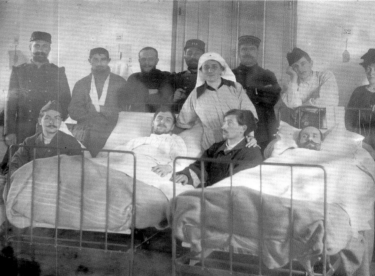
Jeanne Le Moigne nursing wounded soldiers, Pont l'Abbé, 1916.

was called *l'École d'Ouvriers et Contremaîtres* (School for Workers and Foremen) and only officially became one of the *Arts et Métiers* schools in 1901. However, its standards were equally high and its pupils were retrospectively granted the privilege of calling themselves *Arts et Métiers* graduates. The aim of the schools was to turn out young men with sufficient technical knowledge that they would be capable of rising to be foremen in any branch of mechanical engineering they chose to pursue. Rod Day, who has researched the careers of graduates from these schools in the French automobile industry before World War II, wrote:

> The students at the schools, or gadzarts as they were called (gars des arts), were frequently of modest origin, the sons of skilled workers, artisans, technicians, and workshop owners. They were sought after by industrialists in the automobile industry because of their practical training, their capacity for hard work and problem solving, and their long association with the machine and metalworking trades. In the schools they received a solid grounding in mechanics, physics and chemistry, applied mathematics, drafting, and machine and metal shop work.[10]

Day found that more than half the graduates from the schools ended their careers in senior management positions and cites, as examples, Émile Delahaye, Henri Brasier and Louis Delage. Each of these eventually founded his own motor car manufacturing factory, gaining fame which belied their modest beginnings.

---

10. C. Rod Day, 'Careers of Graduates of the Écoles d'Arts et Metiers in the French Automobile Industry, 1880–1940', *Canadian Journal of History*, August 1994.

Chapter One

Although ten or fifteen years after leaving Coatalen would be held up as another example of a 'Gadzart' graduate success, his time at Cluny seems to have been somewhat less glorious. It appears from the records that he joined in the second year of the three-year course and despite the picture of serious hard work portrayed by the Sunbeam biographer Massac Buist, his final exam results were not good and according to the school archivist his file bears the note that when he graduated in 1898 he was deprived of his certificate as a qualified engineer because of bad behaviour![11] What mischief he had been up to is not recorded but none of this seems to have depressed his enterprising spirit or affected his career, as his determination, charm and ability more than made up for the lack of a certificate.

Early Career

After leaving Cluny, Coatalen worked in the drawing offices of a number of motor car makers. Reports vary about the number of firms he worked for and the order in which he was employed by them. In an article about him published in the *Auto-Motor Journal* in 1911 it was stated that he had worked for no less than five firms – de Dion Bouton, Vinot et Deguingand, Clément, Panhard et Levassor and Darracq.[12] In view of the fact that he graduated from Cluny in 1898 and did his military service in 1900–01, it seems improbable that he can have worked for so many companies in so little time. The *Souvenir of Sunbeam Service* booklet (printed 1919) briefly records Coatalen's early career and mentions that it included spells at Panhard, Clément and de Dion-Bouton. An article published slightly earlier in New Zealand gives the same firms but in reverse order, with his first job being de Dion, whilst H.O. Duncan, who wrote about Coatalen in the mid-1920s, only mentions de Dion. Charles Faroux, a friend of Coatalen's for many years, who might therefore be considered a fairly reliable source, gave a speech in 1953 in which he said that Coatalen worked initially in the drawing office of Panhard and subsequently in that of Bayard-Clément, and this seems most likely to be closest to the truth. Some of the confusion is no doubt due to the fact that, at this period between 1898 and 1900, Adolphe Clément was not only a director and major shareholder of Panhard but also constructed and sold Panhard-Clément cars and Clément-de Dion cars from his own works at Levallois-Perret. In view of Coatalen's subsequent career it is interesting to note that Clément coincidentally also had an interest in the Clément-Gladiator-Humber bicycle company and within a few years would set up the Clément-Talbot automobile works in London.[13]

Whatever the truth, it appears it was not unusual for the *Gadzarts* to move around. They had a reputation for having *la bougeotte*

---

11. 16 November 1998. Conversation with M. Hetzlen, Résponsable du Musée de l'Ecole Nationale Supérieur d'Arts et Métiers, Cluny. Coatalen's *'moyenne générale était de 11.76. Il avait son Diplôme d'Ancien Elève de l'ENSAM mais était privé de Diplôme d'Ingénieur pour mauvaise conduite'*. To obtain the engineering diploma, an average score of more than 13 was needed so he only received a diploma to show he was an alumnus of the establishment.
12. *Unconventional Portraits of the Week – Mr L. Coatalen*, the *Auto-Motor Journal*, 2 December 1911.
13. Gerard Hartmann, *Clément Bayard Pionnier Industriel*, ETAI, 2013.

– changing jobs frequently in the early years to advance their careers![14] The story of Coatalen's departure from the Clément works illustrates well his determination to better himself. Having worked there for a while (Faroux says two years), perhaps training alongside the renowned engineer Marius Barbarou who was three years older, he asked for a meeting with the boss himself, Adolphe Clément. At this meeting he requested an increase in pay from the Fr. 150 per month he then received because the cost of meals at the local Bistrot had gone up from 21 to 23 sous. Clément declined, saying that although he probably should and could give him an increase, it would annoy him to have to do it. Instead he offered Coatalen the privilege of using the front staircase rather than the back staircase when entering and leaving the building. Coatalen burst out laughing and, having understood that he was not likely to make fast progress where he was, decided to leave. According to Faroux he was on his way to England within three weeks.[15]

Military Service

We know that Coatalen started his military service in November 1900 (his profession on the military record is given as *dessinateur* – draughtsman) and by September 1901 had obtained *'le brevet de vélocipédiste'* (cyclist's certificate) for which his earlier cycle racing exploits were no doubt a great help. At the end

Louis Coatalen in uniform with bayonet fixed during his military service in 1901.

of the year he was released from active service with a *'Certificat de Bonne Conduite'* (good conduct) and his name was added to the list of reservists.[16] One can only assume that he returned briefly to Clément at the end of 1901 before deciding to try his luck in England.

---

14. See C. Rod Day, 'Careers of graduates of the Ecoles d'Arts et Metiers...'
15. Charles Faroux, *Journal de la S.I.A.*, July 1953, p. 207.
16. Coatalen served with the 118eme Régiment d'Infanterie, Quimper, and was then listed as reservist for the 6eme Régt d'Infie Colonial, Brest. His brother François, as the eldest child of an orphaned family, was officially not required to do military service but he volunteered to do one year's service and thus Louis was only required to serve for a year of the normal two-year service.

Two

Coventry Calls 1901–08

News of openings for engineers in England must have filtered through to Paris to encourage Louis Coatalen to drop everything and set out at the beginning of the new century to seek 'fame and fortune' in a country where they spoke a language of which, as far as we know, his only knowledge came from lessons given to him by the painter Fromuth when he was a child. There may have been other factors that affected his decision to make his move at that moment, but these are merely speculation. For example, if he had applied for a job with another French manufacturer, they might well have wanted to know what result he had obtained in his final exam at Cluny if they were comparing the qualifications of several candidates. This was much less likely to happen in a foreign country. It is also possible that he may have inherited a sum of money as a result of the sale of his mother's hotel in Concarneau in 1899 and this could have encouraged him to take such a major step. However, his father was still in hospital and presumably his care would have had a first call on any available funds, so Louis may not have received anything from this source to set him on his way. What is known for certain is that Coatalen celebrated his twentieth birthday in September and his uncle Hervé Le Moigne formally stopped being responsible for this headstrong youth on 24 October 1899, having wound-up the guardianship which he had exercised for five years.

At the time Coatalen was living in the Rue Brézin, in the 14th arrondissement of Paris, not far from the Gare Montparnasse, an area with a strong Breton population. We do not know the exact timing of his move to England, but it was at the end of 1901 or early in 1902 rather than in 1900 (as had previously been thought). Once he had finished his military service, Coatalen set out on an adventure that would eventually bring him both fame and fortune. The reason he would give later as an explanation was that he felt 'he was a decade too late to stand the best chance of scoring personal success in the French automobile industry' as things were so much further developed in France.¹

Charles T. Crowden (1859–1922)

According to H.O. Duncan, when Coatalen first got to England he worked for a few months with Charles T. Crowden of Leamington. Crowden was a significant pioneer in the early days of automobilism and *The Autocar*'s obituary described him as possessing 'mechanical ingenuity and ability amounting to genius'.² He had designed and produced in the 1880s, in conjunction with Edward Butler, a tricycle, which was the first vehicle in England to be propelled by internal combustion. Crowden had been Chief Engineer at the Humber cycle works where he patented a system for hydraulically pressing the tubes of a bicycle frame into the lugs without the use of brazing, but he had resigned from Humber in 1896 to take up the position of Chief Engineer to the Great Horseless Carriage Co. in Coventry and was amongst those who took part in the original London to Brighton Emancipation run on a Daimler-powered Panhard et Levassor. The Great Horseless Carriage Co., set up that year by Harry J. Lawson, the indefatigable promoter of motor companies, shared premises with Daimler, another Lawson firm, at the Motor Mills, which was a converted cotton mill in Coventry. Both firms were intended to manufacture motor vehicles under licence from Lawson's British Motor Syndicate which held patent rights from the leading continental manufacturers but, in the words of Bunty Scott-Moncrieff, 'they did not succeed in producing motor-cars in any quantity, and such few as they did build ran very badly and broke down constantly!'³ In 1898 Crowden had left the Great Horseless Carriage Co. (which then became the Motor Manufacturing Company) and started his own works in the old H.J. Mulliner coach-building factory at Packington Place, Leamington Spa.⁴ It was here that Coatalen joined him briefly.

After the huge factories of Clément and Panhard, Crowden's works in Leamington must have seemed something of a let down. *The Autocar*'s correspondent recorded a visit to the Panhard works in 1901, noting 850 men at work, 350 lathes, polishing and planing machines, eighty-seven engines in the fitting shop waiting to be dropped into chassis and everything organised with military order and system.⁵ Panhard occupied 30,000m² and produced 500 cars per year, while the 6,000m² Clément works employed 300 workers and also made 500 voiturettes.⁶ Yet, at the same time, it must have been evident from the contrast in scale and professionalism what potential the English market offered for the fledgling industry

1. Massac Buist, *A Souvenir of Sunbeam Service*, 1919, p. 14.
2. *The Autocar*, 17 November 1922, p. 1079.
3. David Scott-Moncrieff, *Veteran & Edwardian Motor Cars*, Batsford, 1955, p. 39.
4. Damien Kimberley, *Coventry's Motorcar Heritage*, The History Press, 2012, p. 150.
5. *The Autocar*, 13 April 1901.
6. Pierre Souvestre, *Histoire de l'Automobile*, Dunod, Paris, 1907, p. 416.

Chapter Two

to catch up and the openings that this would give to a bright engineer. It seems unlikely that Crowden produced many vehicles but one is preserved in the Coventry Motor Museum that dates from this period and therefore gives an idea of the vehicles on which Coatalen would have worked. It was still very much a horseless carriage and not yet an early motor car. Around the same time Crowden was producing steam-powered, self-propelled fire engine conversions and giving lectures to the Coventry Technical Engineering Society on valve gear. In June 1901 Crowden had an accident which 'temporarily incapacitated him from business'[7] and possibly this was the reason he required the help of the young Frenchman at the end of that year.

Humber

Coatalen's command of the English language must have improved rapidly, for he soon moved on to work for Humber. Charles Crowden, who himself had worked for Humber, would have been well aware of the changes taking place there and was perhaps instrumental in helping the young Frenchman to grab the opportunity presented by the new start the firm was making.

The Humber name had been well established in the world of quality bicycles but the firm suffered a number of vicissitudes at the end of the nineteenth century that put it into great difficulty. As A.B. Demaus and J.C. Tarring record in *The Humber Story*, the Coventry Works was burnt down by a major fire in July 1896 and couple of years later, as a result of dubious financial entanglements, all the

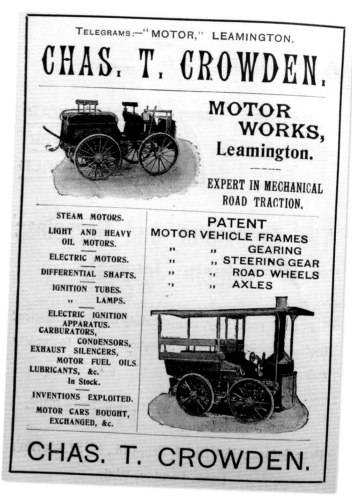

An advertisement by Charles Crowden published in 1900.

directors were dismissed and a new board was formed. An additional complication was added by the diversification of this bicycle maker into the new world of motorised transport when it acquired licences from H.J. Lawson's British Motor Syndicate to build cars under the impractical patents of the eccentric American inventor E.J. Pennington. In order to make a clean start the company was re-formed in 1900 as Humber Ltd., with Edward A. Powell, a

---

7. *The Autocar*, 29 June 1901, p. 637.

solicitor by profession, as the new Chairman. The bicycle range was rationalised and two Wolverhampton factories were closed down, along with a number of overseas subsidiaries.

So, when Coatalen joined Humber in Coventry it had recently been reorganised and rationalised after a financial disaster and was seeking to establish itself as a proper motor car manufacturer. By early 1902 Coatalen had designed the car that was to allow the company to start to build itself up in a serious way. This was a 12 hp car with a four-cylinder engine, four-speed gearbox and shaft drive. Interestingly, it had a patented triangulated tubular chassis frame that was based on the firm's bicycle frame-making experience and technology. The radiator at the front of the car was hinged so that it could be folded down to gain access to the engine. Another unusual design feature, that many would assume was first introduced on the 1955 Citroen DS, was a steering wheel that had only one arm connecting the rim to its centre, but this was already a Humber tradition in 1901.

There is a story that has been repeated down the years, which must relate to the design of the 12 hp Humber and gives an early glimpse of Coatalen's technique for achieving his own ends. This was at a time when it was thought that single-cylinder engines caused enough problems and to multiply the number of cylinders would simply multiply the number of breakdowns. The directors were very conservative but finally agreed to a two-cylinder design. Coatalen produced a two-cylinder casting but then proceeded to mount a pair of cylinder blocks on a crankcase. H.O. Duncan wrote:

> When the day came to inspect the new model, the bonnet was raised and, behold, there were four cylinders! 'Oh, Mr Coatalen,' exclaimed the directors, aghast, 'It's got four!!' Coatalen looked in and, with a well affected air of surprise, said: 'An' so 'e 'ave!'[8]

The Autocar reported enthusiastically on the new car, saying it 'is a most promising vehicle and one which does the greatest credit to its designers and manufacturers and it bids fair to make as great a name for the firm in the motor world as they already enjoy in the cycle world'. The finish of the engine in copper and burnished steel, with its ribbed polished aluminium cylinder heads picked out in red, attracted comment, too, as being 'very handsome indeed ... one of the best looking engines we have inspected'.[9]

Two of these cars were entered for the Automobile Club's 1902 650-mile Reliability Trial, and although one broke down the other acquitted itself honourably. It is perhaps significant that Coatalen's first design was put to the test in this way, in line with his later declared belief that 'racing improves the breed'. A month after this event it was announced that vibration problems encountered at high revs were being

---

8. H.O. Duncan, *The World on Wheels*, Vol. II, p. 900. Duncan tells the story as though it occurred when Coatalen was at Sunbeam, which cannot be correct as all Sunbeams by that time already had four-cylinder engines. Charles Faroux retold the story many years later as having occurred soon after Coatalen joined Humber, which seems much more likely. See *Journal de la S.I.A.*, July 1953.
9. *The Autocar*, 12 April 1902, p. 358. See also *Motor Cycling*, 23 April 1902, p. 183.

Chapter Two

A young Louis Coatalen riding as mechanic to a more elderly driver of a 12 hp Humber.

entirely eliminated by replacing the two-throw crank with a four-throw crank. At the company's Annual General Meeting in November, Mr Powell, the Chairman, reported that 'they now had every facility for manufacturing good motor cars ... (and) they looked upon the motor department as the most promising for future development'.[10] Matters evidently took a while to improve. At the AGM to consider accounts for the year ended 31 August 1904, the depression in the cycle trade and losses in the 'large motor car department' had reduced the reported profit. However, in the 'light-car department considerable sales were effected and a substantial profit realised' so the directors remained very confident for the future.[11]

At this stage Louis did not drive himself, being just the designer. However, by 1905, after the resignation of his boss, the Works Manager, Henry Belcher, who had driven Humbers in a number of competitions, Louis took the wheel for the firm, having grown a moustache again to match his new-found responsibilities.[12] He was perhaps encouraged to drive by the new General Manager of Humber's Beeston factory, Thomas C. Pullinger, who was a firm believer in promoting cars by taking part in events. He often competed himself and was also the man responsible for entering C.H. Cooper and J. Reid in the 1905 Tourist Trophy on the Isle of Man. Having worked for ten years in France for a variety of firms, including Darracq (when it was still Gladiator), before returning to England to join Sunbeam in 1902 and then Humber in 1905, Pullinger was no doubt a Francophile. He would have related well to the enthusiastic Frenchman who was thirteen years his junior and may well have incited Coatalen to promote the Coventry works in a similar manner to his own efforts with Humber's Beeston products. Louis' natural inclinations were already clear, as Humber's Chairman is reported to have said he was delighted to have secured 'the services of that young French boy, but we cannot keep him at his desk in the drawing office; he will get out on the road and drive a car'.[13]

---

10. *The Auto-Motor Journal*, 22 November 1902, p. 784.
11. *The Auto-Motor Journal*, 12 November 1904, p. 1342.
12. Belcher's departure appears to have been temporary, as the following year he is to be found competing again at the wheel of a Humber.
13. Massac Buist, *The History & Development of the Sunbeam Car 1899–1924*, p. 25.

Louis Coatalen's framed pictures recording his non-stop drive from London to Coventry in top gear.

It was the 8-10 hp Humber with which Coatalen is most intimately associated and it was in connection with this car that he first emerged from the shadows and his name started to be recognised both as a competitor and as a designer. The first of these cars was built towards the end of 1904 and *The Motor* noted in its report of the Manchester Show held in February 1905 that the Humber 'is really one of the features of the exhibition and attracts much attention'. It was powered by a four-cylinder, 80 mm x 95 mm engine with mechanically operated valves and a high-tension magneto. Instead of the more usual copper pipes for water circulation, a neat gunmetal casting topped off the water space, giving 'a very smart appearance to the upper part of the engine'. An even more

Louis Coatalen at the wheel of the 8-10 hp Humber about to leave Regent Street on 29 April 1905 on his non-stop run from London to Coventry in top gear.

unusual feature was that the crankcase was made of two cylindrical castings joined vertically at the centre. The crankshaft was supported by phosphor bronze end bearings and a central pedestal bearing. Both crankshaft and connecting rods were steel stampings. The chassis frame was tubular with a 6 ft 8 in wheelbase, 4 ft track and artillery wheels. *The Autocar* commented: 'it may be observed that there is no attempt to put superfluous work on any part of the car, it being maintained that such is unnecessary and only intends to increase the cost, whereas the object of the firm is to provide a fairly cheap and reliable car which shall have a long life and give every satisfaction in working'.[14] By March 1905 the Coventry works were making twelve cars per week of this model. In April 1905 one of them took part in the gruelling, four-day Scottish Reliability Trial but stopped on the final day with carburation

---

14. *The Autocar*, 25 March 1905, p. 423.

problems (T.C. Pullinger's 16-20 hp Beeston Humber completed the trial).

On 29 April 1905 Coatalen drove one of the new 8-10 hp Humbers from London to Coventry using only top gear all the way. On this journey he was accompanied by H. Walter Staner, the editor of *The Autocar* to ensure that this demonstration of his faith in his new model attracted adequate publicity. This seems to be the first time Coatalen's name was publicly acknowledged as the person responsible for the design of the Humber, and his photograph and name appear in the magazine. It is noteworthy that just a few months before, a 30 hp Napier had made the same journey and much had been made of the wonderful flexibility of its six-cylinder engine that enabled it to pull top gear all the way. Here was Coatalen with his modest four-cylinder 8-10 hp car demonstrating it could perform just as well. Coatalen and *The Autocar*'s editor started from the Regent Street garage in the early hours of the morning to avoid traffic and tackled the 'long incline up to Tally Ho corner', then Barnet Hill, the hill into St Albans, and later the hills through Weedon and Dunchurch, all with the greatest of ease.

The 8-10 hp Humber on arrival in Coventry, rain-soaked and mud-spattered some three and a half hours later.

Chapter Two

He purposely slowed at the foot of some hills to 5 or 6 miles an hour in order to demonstrate how speed could be quickly picked up again. It cannot have been a pleasant journey, as for some 60 of the 90 miles 'heavy rain fell without intermission' and they were thoroughly soaked by the time they reached their destination.[15]

In June Coatalen took part in the South Harting Hill Climb and won his class, as did Pullinger with the larger model. In July at the Brighton Speed Trials Coatalen won a scratch race for cars priced between £200 and £350, and then at the Blackpool Race Meeting he won another series of sprints for the Lancashire Handicap along the mile and a half of the new esplanade that had just been completed between the sea and the old footwalk. He also took part in the very first Shelsley Walsh Hill Climb, where he was placed fourth on handicap, recording a time of 2 minutes 20⅗[th] seconds, which was within the top twelve times of the day. In early September there were sprints over a mile on Skegness Sands, where Coatalen won a couple of races and some weeks after that he again drove an 8-10 hp Humber in the 3-mile hill climb up Snake Hill, near Glossop in Derbyshire. In December 1905 and January 1906 Humber's Lincolnshire agent, R.M. Wright, put a 10-12 hp car through a 5,000-mile trial to demonstrate its reliability and ability to withstand adverse conditions. On two days when Wright was unwell, Louis Coatalen took over the driving himself.

Coatalen demonstrated that even in his very first year of competitive motoring he was a shrewd judge of what he could win and, by sticking to his small 8-10 hp car, made it clear that both he and it were a force to be reckoned with. He admitted that the car he drove at Brighton was much more highly geared than standard, as it had exceeded 50 mph, but to demonstrate its flexibility he accepted the challenge to climb Birdlip Hill in Gloucestershire with three passengers on board. This he managed using the same car (he took 4 minutes and 20 seconds) and won the wager.

It is indicative of Coatalen's position at this stage in his career that he did not appear as the entrant or owner (in contrast to T.C. Pullinger) of these cars but simply as the driver – mostly at the wheel of the car belonging to Mr J.W. Adams (before the latter died in April 1906).[16] Coatalen probably did not own a car at that time as, according to Hugh Rose, he insisted on

At the Brighton Speed Trials in 1905. Coatalen competed at the wheel of Mr J.W. Adams' 8-10 hp Humber.

---

15. *The Autocar*, 6 May 1905, p. 618.
16. J.W. Adams was Sales Manager for Humber. A couple of patents were taken out jointly in his and Humber Ltd's names. Before the 1905 TT Race, Adams took *The Autocar*'s correspondent around the course in one of the 16 hp Humbers entered for the race.

The engine of the 10-12 hp Humber. Note the vertically divided crankcase.

wearing his bicycle clips all day and every day, whatever the occasion.[17]

His importance within the company, and the trade generally, blossomed in 1906. Coatalen took part in a couple of hill climbs with the upgraded 10 hp Coventry Humber but the main concentration of effort went in preparing a special 20 hp Coventry Humber for the Tourist Trophy Race in September. This was to be his debut in serious motor racing and also marked his emergence as one of the sport's personalities. In early July we find *The Autocar* recording the comments of 'M. L. Coatalen, the designer and driver of the Coventry Humber car entered for the Tourist Trophy Race' who had just returned from a tour of inspection of the Isle of Man course. He is reported to have said that many of the roads were 'in a shocking state' and 'exceedingly bumpy' and that unless improvements were made the race would 'be a much more severe trial … than last year's event'.[18] Then, at the beginning of August, the magazine published a photograph of the Tourist Trophy Coventry Humber 'with its designer M. Coatalen at the wheel'. He looks very dapper in a suit, collar and tie and wearing a cap, but it is not possible to know whether he is wearing his bicycle clips.

When the reporter for *The Autocar* arrived on the Isle of Man before the race, he wanted to explore the course and 'within an hour of our arrival we were fortunate to meet

---

17. Barrie Price, *The Lea-Francis Story*, Batsford, 1978, p. 105. I am grateful to Allan Lupton for drawing my attention to this reference.
18. *The Autocar*, 7 July 1906, p. 30.

Chapter Two

Coatalen and the 1906 Tourist Trophy Humber.

Mr Coatalen ... he immediately proffered the use of the Humber second car which we gratefully accepted'.[19] Subsequently, Coatalen took the reporter for a lap in the car he was going to drive in the race. The reporter wrote:

> Up to a certain point the fates appear to have been against the Humber, but we are pleased to say that fortune at last appears to be smiling upon the efforts of Mr Coatalen to bring his car up to the exact concert pitch. Owing to damage caused by the careless handling of one of the cylinders of the engine, we were not able to make a run on this car until Tuesday morning last, but then the circuit was completed in excellent time, and what is more, a little fuel was left at the finish. The start was made from the official point, previous to which the petrol tank had been emptied, and the regulation quantity placed therein. We might go on to say that we were distinctly favourably impressed with the running of the Humber car. It sits on the road more like one of the heavy cars, and depressions and rises in the road do not affect it to the same extent they usually do on medium weight cars. The springing and the upholstering of

---

19. *The Autocar*, 22 September 1906, p. 400. This 'proved to be the standard 10-12 hp Coventry Humber driven so successfully by Mr Wright of Lincoln in several trials and competitions since its 5,000 mile demonstration'.

the vehicle are certainly good, and tend to very easy riding, even at the fastest speed. From Douglas to Peel we noticed that the needle of the speedometer never descended below thirty miles per hour, and, although a strong headwind was blowing against us, a steady thirty-five was given. Away from Peel to Kirkmichael, Ballagh, and Sulby, we were well on the way to forty-five, and fast time was done along this piece.

The mountain climb from Ramsey was taken in very good style until we omitted to keep up the pressure in the petrol tank, when the engine faltered and stopped. The Humber feed to the carburettor has up to the present been by gravity, but it has been found that this is rather inclined to starve the carburettor on very long rises, and therefore, as a temporary measure, pressure feed was fitted up with a view to testing, and if found satisfactory, the official type of pressure feed float, etc., was to be fitted. On getting pressure in the tank again, the engine started up well, and pulled magnificently to the summit of the mountain climb, after which, the nature of the road allows of an exceedingly high speed to be taken with comparative safety. Nearing Douglas on the road above Keppel Gate, the engine was stopped, and a coast down at about forty-five miles per hour, and rounding the corner at the Keppel hotel, a further long coast resulted, the speed obtained at some points being well over fifty miles per hour. Of course only sweet-running parts and certain acting brakes would enable such high speeds to be obtained with a touring car, and we must say that during the whole of this run, never once had we any feeling that there was any danger whatever, however fast we went. On arriving at the finishing point, the question with us was, had we, or had we not, done the course on the consumption, for to be quite sure that we should get the proper feed to the carburettor on gravity when on slight rises only, we had put an extra two gallons of petrol in the tank, and, of course, this had to be taken out at the finish. In doing this we found that about a quarter of a pint of petrol was left, at which Mr Coatalen rejoiced exceedingly. The time for the course was 1h 14½m., a fairly fast performance, although not as fast as some of the cars, but the fact that the petrol consumption was all right is one that should give the Humber car an excellent chance in the race next Thursday.[20]

A notable detail is that Humber's characteristic tubular chassis frame was replaced with a more conventional girder type chassis that was extensively lightened with a latticework of holes. Despite this, Coatalen's car was nearly 4 cwt heavier than the Beeston Humber,

---

20. *The Autocar*, 22 September 1906, p. 401.

which must have made the 25 miles per gallon consumption demanded by the regulations particularly challenging to achieve, especially as we know that Pullinger, who drove the Beeston Humber, had great difficulty in getting near to this performance.[21] All-out speed was out of the question. As revealed by *The Autocar*, once Coatalen got to the top of Snaefell the engine was turned off and gravity took over for the descent. With this technique he managed to finish sixth, but even so, after five hours of racing, there was only a very small amount of petrol left in the bottom of the tank at the finish. Despite the reassurances of *The Autocar*'s reporter, it was evidently a technique that involved considerable risk as Coatalen, who sounds as though he was beginning to enjoy his notoriety, was quoted in the local press as saying:

> Eef there is a smash I tink certainly some of zem will go downwards. But I am all right. I haf a very good boy for my mechanician. He never swear, but always go to church on Sunday; so I say to him, 'If we haf a smash I shall hang on to you – you must introduce me!'[22]

That mechanic was one F. Howarth. Fortunately he was not required to put in a good word with God as they crossed the finishing line intact. On the first of the four laps of the race, dirt got into the carburettor, which hindered Coatalen's performance and may explain why he was timed over a mile up the mountain road at just 16 mph compared to Charles Rolls, in the winning Rolls-Royce car, who went up at 26 mph. However, Coatalen's Coventry Humber was slightly slower than Pullinger's Beeston Humber throughout the race. Their average speeds, for the five hours that the race lasted, were 32.1 mph (Coatalen) and 32.7 mph (Pullinger). Both were consistently outpaced by the speed of the winning Rolls-Royce, which averaged 38.1 mph. Pullinger, who finished just in front of Louis, declared after that race he was never going to take part in a competition again! However, they both had the satisfaction being amongst only nine finishers out of twenty-nine starters and the only team to finish, which was a great improvement on the previous year's result when neither Humber reached the chequered flag.

This was evidently the beginning of the end of Coatalen's time with Humber and it is remarkable how, during the time that he worked there, he had established himself as a known and respected figure in the motoring business. Within five years he had worked himself up to a situation where *The Autocar* referred to him as having 'certainly established himself as a successful designer' and now he was also known as a competitive driver as well. How much of Humber's success at the time is due to Coatalen's design work and how much to the good management of others is hard to judge but it is on record that the company's profits by the end of Coatalen's time had risen to substantial figures that were not to be reached again for many years. Profits in 1904 were £1,225, in

---

21. *The Autocar*, 06 October 1906, p. 474.
22. Cited in Demaus and Tarring, *The Humber Story*, Alan Sutton, 1989, p. 61.

Crossing the finishing line at the 1906 Tourist Trophy Race on the Isle of Man. Coatalen finished in sixth place, having averaged 32.1 mph for five hours.

1905 £6,537, in 1906 they leapt to £106,558 and in 1907 they reached £154,537.[23] Coatalen's Humber designs became so popular that, according to the chairman, Edward Powell, 'We were almost driven to assembling the vehicles in the streets so greatly did the demand so outstrip our factory capacity.'[24] After Coatalen's departure the company was hit by the recession, which resulted in a restructuring in 1908 and in early 1909 both the Chairman, Mr Powell, and T.C. Pullinger resigned. In Edward Powell's case this was only temporary, as he continued to be Chairman for many years after that.

---

23. Demaus and Tarring, *The Humber Story*. By comparison, Sunbeam M.C.C. profit for 1904–05 was £3,808 while Rover made £16,211 and Darracq £152,664.
24. Cited in *NZ Motor & Cycle Journal*, 25 October 1918, p. 105.

Private Life

Louis Coatalen had wasted no time in putting down roots when he moved to England. It is known that Humber organised opportunities for social interaction amongst employees, such as an annual fancy dress ball, but it is not known whether this was how Coatalen met his first wife. On 24 November 1902 he married Annie Ellen Davis in Birmingham. Although a son was born in August 1905, who was given the names Louis Hervé, it is thought that Louis and Annie were separated some six months after they married. They were divorced sometime in 1906 and Annie remarried in 1908. Certainly this first marriage was not a success and according to his daughter, Marjolie Coatalen, 'my father always claimed that he had no recollection of this first wife, and could not even remember her first name. This caused him no end of trouble with the French bureaucracy on the occasion of his subsequent marriages!'[1] In 1904, according to his military record, Coatalen was living at 47 Alfred Road, Coventry but in 1905 moved to 13 Newport Road, Balsall Heath, Birmingham. After his father died, Coatalen revisited his hometown of Concarneau in July 1906 but it seems he went with some colleagues rather than his wife. When he called on Charles Fromuth, the painter who had taught him English as a child, Fromuth noted in his diary that Coatalen 'has become a celebrity in automobile construction in Coventry England and visited me with three personages in his affair'.[2] As we have seen, 1906 was certainly the year that his fame really spread and it was true to say that by then he was a celebrity. He was still only twenty-seven years old.

Hugh Rose, who was an apprentice at Humber during Coatalen's time there, said that Coatalen advised him that the best way to advance his career was by marrying the chairman's daughter. There was evidently little scope for this within Humber but William Hillman had six daughters and also wanted to set himself up as an automobile manufacturer. According to Rose, this was one of the reasons Coatalen left Humber at the end of 1906.[3]

---

1. Marjolie Coatalen in her memoirs states that Louis and Annie had been separated soon after their marriage and long before the birth of Louis Hervé Coatalen junior and that somebody else was the father. Annie Davis married Commander Horatio Walcott Colomb, son of a Vice Admiral, in 1908. He had retired from military service after the Boxer Rebellion and worked as an insurance agent. Annie was his second wife and he was some twelve years older than her.
2. Charles Henry Fromuth, diary entry for July 1906 after Louis Coatalen had been to visit him. My thanks to Françoise and Jean-Michel Gloux of Concarneau for allowing me access to Fromuth's diaries.
3. Barrie Price, *The Lea-Francis Story*, Batsford, 1978, p. 105.

Hillman-Coatalen

At the beginning of March 1907 *The Autocar* reported that a Hillman-Coatalen car had been entered for the Tourist Trophy Race to be held at the end of May. The formation of this new partnership to manufacture motor cars by two men, who were described as 'a sound mechanical engineer and a talented designer', was evidently quite recent and although a race entry had already been submitted the car at that stage was still only a design on paper. 'We have seen the designs and we must say that we think M. L. Coatalen, who has certainly established himself as a successful designer, has excelled himself in his latest design.' This is indicative not only of Coatalen's priorities and the importance he placed on competition as a means of promoting his products but also of his confidence in his own ability. He entered himself to drive his new untested model in a gruelling 240-mile race whilst it was still just an idea. He was no longer just a designer but a part owner of the business, with his name on the radiator of the cars he was creating. He had really made his mark in Britain.

William Hillman (1848–1921), 'the sound mechanical engineer', who was thirty years older than Coatalen, had made his fortune as a manufacturer of bicycles at the end of the nineteenth century. He no doubt financed this move into the world of the automobile, emulating a number of his contemporaries who had already taken a similar step. By early May the first Hillman-Coatalen car was on the road and *The Autocar* published a photograph of the two partners sitting proudly in their four-cylinder, 20 hp car. Power was transmitted through a three-speed gearbox to a live rear axle, but there appears to have been only one car produced at that stage, with little back-up available for Coatalen's entry in the race. For the Tourist Trophy Race his riding mechanic was a local man, Jim Broadbent, who was only engaged on the Isle of Man itself. The cars were filled with the allotted amount of fuel (9 gallons 6 pints) the night before the race and then locked away. Coatalen was horrified to find the next morning that a slow leak had been dripping all night. It was evident that he had no hope of finishing, so he set out with the aim of putting in the fastest lap. This he did in fine style despite the appalling weather, overtaking many competitors who had set off before him. He completed the first lap in the lead in 1 hour, 4 minutes and 39 seconds, an average of 37.4 mph – nine minutes faster than the Beeston Humber which was the next most rapid. On the second lap he had what was described as 'a slight dispute with Quarter Bridge', damaging a rear spring which later broke and caused him to retire. As a result it was never clear how much petrol had been lost or how far he might have gone in the race. A repair must

Louis Coatalen and William Hillman in the new Hillman-Coatalen car in May 1907.

Chapter Two

The Hillman-Coatalen 25 hp engine.

have been effected, as the following day Coatalen took part in the Graphic Trophy Hill Climb but was outclassed by more powerful vehicles. A week or so later he was back on the mainland taking part in the Henry Edmonds Trophy Hill Climb contest, which included a stop and restart element, but he only came in eighth.

Perhaps after this Coatalen decided more development work was needed, as he was notably absent from the second-ever meeting to be held at the new Brooklands race track in July that year, which featured a race for Tourist Trophy cars. He was not alone in being absent, however, as the race only attracted three entries. It is interesting to note that the Board Meeting of the Hillman-Coatalen Company, at which Louis was appointed Managing Director and allotted 10,000 shares in the business, took place on 17 June 1907 at 38 Bailey Lane, Coventry, long after the prototype's first appearance and the TT race. At the same meeting William Hillman and Harold Smith also took office as Directors of the Company. The next mention of the Hillman-Coatalen car occurs right at the end of August, by which time it was said to be ready for the road, although a full description was not published until October. By the time of the Motor Show in November, a 'still somewhat incomplete' six-cylinder model was on display beside the four-cylinder, 25 hp car which was described in detail. Cylinders were cast individually, having a 5-in (125-mm) bore and stroke, with inlet valves on the left and exhaust valves on the right. The two-piece, barrel crankcase was bolted together at the centre and the crankshaft was carried on three bearings. Although rated at 25 hp it was thought that it would undoubtedly produce nearer 60 hp on the brake. It had a three-speed gearbox with direct drive in top through a live axle and the whole car was described as reasonably priced, light yet strong. In view of Coatalen's absence from competition at this period, one has to assume that most of 1907 was taken up with getting the new factory set up and the car ready for production. By October it was reported that 'the newly built and equipped works at Pinley, near Coventry, are now in full swing'.[25]

In 1908, although Coatalen's road car was a very raffish looking racer that was even referred to as the 'Hillman-Coatalen Brooklands racer', it does not appear to have raced at all. It was

Louis Coatalen and his mechanic pose before the 1908 TT Race in the Hillman-Coatalen, in which they finished in ninth place.

25. *The Autocar*, 12 October 1907.

Chapter Two

Kenelm Lee Guinness in the Hillman-Coatalen for the 1908 Tourist Trophy Race. He retired with a broken chassis.

a stripped version of the 25 hp model with 5-in bore and stroke, but with a five-bearing crankshaft, two water pumps, a Bosch magneto and a coil for starting. There is a suggestion that this was the previous year's TT car that had been rebuilt. Both Coatalen and Hillman were seen at Daventry on this car when they were spectators at the 2,000 miles trial.

By August 1908 Coatalen had expanded the range with a smaller 12-15 hp, four-cylinder (3½ in x 3¾ in) engine with the crankcase divided vertically, in 'the normal Hillman-Coatalen manner' (a design feature he had first introduced on the 1905 8-10 hp Humber), which was claimed to give great strength. The three-speed gearbox with gate change drove a live rear axle.

The only competition that Coatalen entered in 1908 was the Isle of Man Tourist Trophy Race. It was the first year that the TT became a real race and no longer an exercise in minimising petrol consumption. The only limitation was that the cylinder bores should not exceed 4 inches. Two 25 hp Hillman-Coatalens were prepared for the race, held in September, which were 'practically standard' except the bores were reduced to 4 in, and 3-in valves were fitted. The induction pipe was oval in form and Rudge Whitworth detachable wire wheels were used. One car was, as before,

Coatalen motoring hard through Cronk-ny-Mona soon after the start in the 1908 Tourist Trophy Race.

for Louis Coatalen, while the second was for Kenelm Lee Guinness to drive. This marked the beginning of a long and close collaboration between the two men. In the race Guinness retired with a broken chassis but Coatalen kept going to the end, covering almost 340 miles in 8 hours and 20 minutes – an average speed of 40 mph. Even though this average for the whole race was well up on his single record-breaking lap speed of the previous year, it was

Chapter Two

just not quick enough when compared to the competition – the winning Hutton averaged 50 mph – and Coatalen came home down in ninth place.

These disappointing competition results cannot have helped matters but it seems clear that the business as a whole was not successful, and this led to it being put into liquidation. In October 1908 one of their suppliers, United Motor Industries Ltd, petitioned for the Hillman-Coatalen Motor Car Co. Ltd to be wound up, having no doubt waited a while for bills to be paid.[26] A Receiver, Mr Geoffrey Bostock, had already been appointed by the beginning of October, and it is on record that early in November 1908 Louis relinquished his directorship and transferred his shareholding back to William Hillman.[27] Then on the 18 November 1908 an Extraordinary General Meeting was held that confirmed 'that the Company be wound up voluntarily'.[28] Coatalen's dream of producing a car under his own name had only officially lasted eighteen months, although the range of three Hillman-Coatalen cars (12-15 hp four-cylinder, 25 hp four-cylinder, and 40 hp six-cylinder) continued to be sold throughout 1909 under that name. Curiously, in a report on 'The State of the Motor Industry in Coventry' published in *The Autocar* in February 1909, the company was reporting 'trade improved wonderfully' and that if it continued they would have to work overtime![29] Somehow the Hillman-Coatalen Motor Car Co. Ltd issued £1,000 of debentures out of an authorised total of £3,000 in the summer of 1909 and it was not until 1910 that the business became the Hillman Motor Company.[30]

---

26. *London Gazette*, 30 October 1908, p. 7850.
27. Information from Sarah Strover, Coventry Transport Museum archives, 2 February 2009.
28. *London Gazette*, 15 December 1908, p. 9590.
29. Hillman-Coatalen cars continued to be sold under that name throughout 1909 despite Coatalen having left.
30. *The Motor*, 7 September 1909.

Private Life

Before leaving the subject of Hillman-Coatalen cars, some mention should be made of William Hillman's daughters (sometimes referred to as the Hillman Minxes), if only to repudiate the oft-repeated myth that Louis Coatalen married one of them. Hillman had six daughters and at least two of them were later to marry significant figures in the motor industry – Edith married Spencer Wilks (eventually Managing Director of Rover) and Margaret married John Black (eventually Managing Director of Standard Triumph); both men had worked for Hillman becoming Joint Managing Directors of the firm. By 1908 Coatalen was twenty-nine years old and there would have been six girls to choose from between seventeen and thirty-two years of age. Coatalen had married and divorced his first wife during his time at Humber and so was single by the time he joined Hillman. He would have theoretically been eligible to be a son-in-law (if perhaps not a very desirable one being a divorcee). It is not known what sort of relationship Louis may have had with any of the Hillman daughters or whether this might have had some sort of influence on the ending of the Hillman-Coatalen partnership. Although it seems quite probable that the relationship with William Hillman may have been strained as a result of Coatalen's interest in

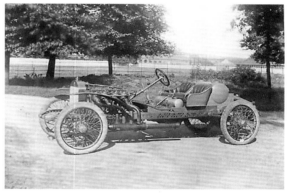

Top: Thought to be Coatalen's personal 25 hp Hillman-Coatalen (picture from an Olive Coatalen album).

Bottom: Louis Coatalen driving a six-cylinder 40 hp Hillman-Coatalen at Harley Bank Hill Climb in October 1909, with his new wife Olive by his side.

the girls, he did not get married during his time with Hillman, which was convenient as this meant he was able to marry Miss Bath soon after joining Sunbeam. William Hillman was described in his obituary as being 'of a very reserved and retiring disposition so that he was not known personally' to many in the trade.[1] It may have simply been that these two very different personalities could not get along.

---

1. *The Motor*, 9 February 1921.

Three

Wolverhampton Wonders 1909–14

Sunbeam 1909–14

The Sunbeam Motor Car Company Ltd of Wolverhampton (SMCC) was set up in 1905 but its origins stretch back to the middle of the nineteenth century. The central figure was John Marston (1836–1918) who started as a manufacturer of tinplate and japanned goods in 1859, aged twenty-three, after serving an apprenticeship in that trade. His pots and pans business flourished and in 1887 he founded the Sunbeamland Cycle Factory, in which the highly renowned bicycles were manufactured with considerable success. From 1892 these top-quality bikes were fitted with the distinctive, patented, Little-Oil-Bath chain guard for which they became famous. In 1898, Marston's son, Charles, opened an 'accessories factory', known as the Villiers Component Co., to supply pedals and other such components for Sunbeam bicycles. Thomas Cureton, who had started as an apprentice to Marston in 1877, aged fifteen, but had soon risen to be his right-hand man, became increasingly aware of the growing potential of the automobile as the end of the century approached and persuaded Marston to allow him to experiment with building a car. Funded by the Marston fortune, various trials were made from 1899 onwards that led to limited production, but it was not until November 1902, after Thomas Pullinger was employed, that Sunbeam decided to become a serious 'manufacturer' of motor cars. Pullinger had recently returned from working in France and put forward proposals that Marston should start to build up a reputation by importing parts from Berliet in Lyon and assembling them in Wolverhampton under the Sunbeam name. Pullinger was a firm believer in entering cars for testing events as a way of demonstrating their capabilities and promoting sales. As the business grew and started to manufacture its own products, a drawing office was established, with B.J. Angus Shaw as designer to put into practice the ideas and tendencies that Pullinger reported were evolving in France.

Pullinger had written in a 1902 report and recommendations to the directors of John Marston Ltd about the Berliet project, 'I thank you for your promise to make me a Director of this new Company, and to give me a certain number of shares in it. I also note that my salary will be increased and that I shall have the management of the works and the proposed Sales Department.'[1] For whatever reason the Sunbeam Motor Car Company was not floated as a public company until March 1905, with Thomas Cureton as Managing Director, by which time Thomas Pullinger had left and moved to join Humber. However, he left behind him a well-established tradition of publicising the cars through a variety of events that was continued enthusiastically by employees such as Frederic Eastmead and Edward Genna. For example, in 1906, at the wheel of an Angus Shaw designed 16/20 Sunbeam, Eastmead undertook a non-stop run from John O'Groats to Land's End and back again (1,756 miles) without once stopping the engine, and in 1907 he did a similar end-to-end run in Ireland.

By 1908, as has been demonstrated in previous chapters, the reputation of the

---

1. Thomas Pullinger, report to directors of John Marston Ltd, 11 November 1902, cited in *Motoring Entente*, p. 18.

Louis Coatalen (designer) at the wheel of the chain-driven 14-18 Sunbeam in July 1909. In the front passenger seat is Thomas Cureton (Managing Director).

Chapter Three

young Frenchman, Louis Coatalen, was well established in British automobile circles. Cureton would have been aware that he was at a loose end when he parted from William Hillman and perceptively saw that his talents could be of use to the Sunbeam Company. Sunbeam's designer, B.J. Angus Shaw,[2] left to join Thomas Pullinger at Arrol-Johnston and Cureton employed Coatalen as Designer and Chief Engineer. It is not known whether the appointment of Coatalen was the reason for Shaw's departure or whether his departure provided the opportunity for Coatalen's appointment. Either way, by February 1909, in the words of Kent Karslake, 'a bombshell hit Wolverhampton'.[3]

1909: Getting Started

Already by mid-March *The Motor* was reporting that the Sunbeam works 'were very busy under Mr Louis Coatalen' and that he was so satisfied with the set-up that exceedingly few alterations were planned, although a new live-axle car would be making its debut in the Scottish Trial. When this car emerged it not only had live-axle drive but was powered by a new four-cylinder (cast in pairs) T-head engine with 95 mm bore and 135 mm stroke.

The Scottish Reliability Trial took place in June with Coatalen himself driving and Henry Bath, a director of the company and soon to become his father-in-law, by his side. Two other

2. *The Auto-Motor Journal*, 2 March 1912. B.J. Angus Shaw was at Sunbeam for seven years, then for three years he was Works Manager for Pullinger's Arrol-Johnston Co., Paisley, before joining the London General Omnibus Co. as Works Manager in 1912.
3. Karslake and Nickols, *Motoring Entente*, Cassell, 1956, p. 41.

Coatalen at the wheel of the 14-20 Sunbeam, accompanied by Henry Bath and an observer during the 1909 Scottish Reliability Trial. Having climbed Cairnwell from Braemar on the highest road in Great Britain, they are seen descending via the Devil's Elbow to Pitlochry on the fifth day of the trial.

Harley Bank Hill Climb, October 1909. Coatalen is at the wheel of a 16-20 Sunbeam, with which he won the Star Trophy.

Sunbeams were also entered by the factory: Edward Genna on a 14-18 hp car and the indefatigable Frederic Eastmead on his famous 20 hp 'Snark' that had already covered 40,000 miles. The course of the trial covered 1,000 miles in six days through beautiful Scottish countryside on some suitably challenging roads. On the first day they drove from Glasgow to Aberdeen, the second day from Aberdeen to Inverness, the third day was a circular run back to Inverness, the fourth day from Inverness to Pitlochry, from where there was another looped route on day five and finally on the sixth day they returned to Glasgow. The new car came through without having to make any involuntary stops to finish second in its class and ahead of its two colleagues. *The Autocar* correspondent remarked upon Coatalen's clever driving and that he was 'justly proud' of the new live-axle Sunbeam that gave a good impression.

This was just the beginning of Coatalen's personal campaign to draw attention to Sunbeam cars. Soon afterwards he sent a power curve to *The Autocar* showing that the engine produced

27 hp at 1,000 rpm, 40 hp at 1,500 rpm, and 50 hp at 2,000 rpm. In July he entered a 14-20 hp for the Shelsley Walsh Hill Climb, although this demonstration was less successful than hoped as the carburettor flooded and he had to stop on the way up to release the needle valve. The following week he was at Brooklands to obtain a RAC Trial Certificate for one of these cars.

Before the end of August details of Coatalen's most important new car were beginning to emerge. This was the 12-16 hp Sunbeam with its four-cylinder (cast in pairs) engine, with valves on opposite sides (the 'T' head), and 80 x 120 mm bore and stroke giving it a capacity of 2.4 litres. In common with a number of other manufacturers at the time, the crankshaft was 'désaxé' by ⅝ in to obtain a more direct thrust on the crankpins and thereby minimise piston friction and bore wear. The rear wheels were driven via a live rear axle. Although it did not go into production until later in the year, it had taken little more than half a year to create the new medium-sized, medium-priced car that was to help to turn around the fortunes of the company. Coatalen, meanwhile, even found time to return for a brief holiday to Loctudy in his native Brittany, accompanied by Olive Bath, the daughter of Henry Bath, whom he then married on 3 September before celebrating his thirtieth birthday on 11 September.

There was a final competitive outing on 2 October 1909, when the local Wolverhampton Automobile Club held a hill climb at Harley Bank, Much Wenlock, on for which no less than four 16-20 hp Sunbeams were entered. In front of a large crowd and in fine weather they were driven by Coatalen, Bayliss, Genna and Cureton, but it was Coatalen who came first in the Club Handicap, winning the Star Trophy as well as the Club Medal for the class. Curiously, in the larger car class, he drove his two-seater, six-cylinder, 40 hp Hillman-Coatalen, which he must have retained from his previous employment. Accompanied by his new bride, he made the second fastest time in class V, taking 2 minutes and 15 seconds, some fifteen seconds quicker than H. Nelson Smith in a similar car.

On 10 November the capabilities of the new 12-16 hp Sunbeam were officially established when it obtained an RAC Monthly Trial Certificate of Performance. In the road test over 105 miles, its average consumption was 20.12 mpg whilst on the track its fastest lap was at 49.34 mph. In the acceleration test it reached 15 mph in 3.6 seconds and 30 mph in 13.96 seconds, and it climbed the test hill at 12.709 mph. By then Sunbeam's model range for 1910 was firmly established; in addition to the 12-16 hp car there would be the 16-20 hp and an even bigger 25-30 hp model. In deference to Sunbeam's tradition of a chain-drive encased in an oil bath that linked the motor cars with their bicycle heritage, both the larger models were to be available with the option of chain-drive or live-axle drive. Apart from that concession, Coatalen had put his imprint very firmly on the company's products before his first year at Sunbeam was over.

During the winter, work on fine-tuning the 12-16 hp car went ahead and a couple of stripped-down versions were prepared for the forthcoming competition season. By now Coatalen was supported in the experimental department and drawing office by a young

Chapter Three

Believed to be the only known photograph of the engine for *Nautilus*, with four overhead valves per cylinder operated by push rods and rockers across the head.

man called Hugh Rose (1886–1965) who had done his apprenticeship at Humber during Coatalen's time there and joined Sunbeam as head draughtsman.[4] Together they embarked on the development of a one-off, single-seater car to be used purely for racing; this was given the name *Nautilus* because of its pointed, cigar-shaped body like that of the fictional submarine conceived by the author Jules Verne who was also a Breton.[5] This was at a period when a number of racing cars built for the Brooklands track were not only experimenting with so-called 'wind-cutting' bodywork but were also often given individual names, so Coatalen was making his distinctive contribution to this trend on both counts. The aluminium bodywork with its brass, cone-shaped nose and tail was mounted on a chassis with oil-bath-enclosed chain final drive. It looked very striking, but the four-cylinder (92 x 160 mm, 4,257 cc) engine under the bonnet was of even more interest. Although engineers by then were aware of the theoretical benefits to combustion chamber shape offered by overhead valves, the best means of driving those valves had not been established and there were few examples to follow. However, Coatalen would undoubtedly have been familiar with the Guinness brothers' 200 hp Land Speed Record Darracq whose V8 engine (designed by Paul Ribeyrolles) had pushrod and rocker operated overhead valves. *Nautilus* was Coatalen's first experiment in using overhead valves and he arranged four vertical valves above each cylinder operated by push rods from camshafts on either side of the block and by long rockers across the head.[6]

Louis Coatalen at the wheel of *Nautilus* in 1910, his first overhead valve engine racing car. This photo shows the full streamlined body, in aluminium with brass nose and tail cones, and the enclosed oil-bath case for the rear driving chains. The rear of the body was later removed to improve cooling. It lapped Brooklands at 77.5 mph in the March Handicap.

---

4. According to Barrie Price, Raymond Hugh Rose, the son of an engraver, born in Southampton, educated in Taunton, was apprenticed to Humber c. 1903 to train as an engineer and left in 1907. He may have followed Coatalen to Hillman before moving to Sunbeam. Another ex-Humber employee to join Sunbeam as Works Manager in June 1912 was Sydney Guy. When Guy left to set up his own lorry manufacturing works in May 1914, Hugh Rose joined him as designer.
5. Frank Bill was the young mechanic who helped to build the car. Letter from Frank Bill to Anthony Heal, 7 February 1943. (author's collection)
6. When Anthony Heal prepared his first article for *Motor Sport* about this car in 1947, he not only corresponded with Hugh Rose but also arranged for Jean Coatalen to transmit a draft copy of the article to his father Louis Coatalen in Paris, whose comments were then relayed back by Hervé Coatalen and incorporated in the article. Anthony finally met Louis Coatalen in Paris in December 1955.

1910: Brooklands Beckons

Louis Coatalen entered both *Nautilus* and the stripped 12-16 hp car for the first Brooklands race meeting of the year where they made an instant impact. He won the Raglan Cup for 16 hp cars by 150 yards, averaging 55.5 mph, at the wheel of the 12-16 hp Sunbeam, putting in an impressive final sprint. *Nautilus* lapped at 77.5 mph in its race, the March Handicap, but had to slow to avoid an accident between two other cars and finished in ninth place. However, although the stunning body of *Nautilus* may have cut through the wind externally, it evidently did not provide sufficient cooling air flow internally and the rear half had to be removed. Even with this modification the car overheated on its next appearance at Brooklands in June and did not reappear thereafter.

In contrast, Coatalen's 12-16 hp (named

Coatalen and *Nautilus* about to start a race at Brooklands.

Toodles, the Bath family nickname for Olive) along with another one driven by N.F. Bayliss, appeared at numerous events throughout the rest of the year. In addition to the three other Brooklands meetings (where between them they clocked up two first, five second and one third place), one or the other or both put up respectable performances in hill climbs and sprint events that included Saltburn Sands, Ironbridge and Shelsley Walsh. At the latter event Coatalen's climb was much improved on his previous year's performance and he finished third in the Henry Edmunds Trophy for 16 hp

Toodles in road-going trim.

Louis Coatalen at Brooklands on *Toodles*, the stripped and tuned 12-16 Sunbeam with which he won the Raglan Cup at the 1910 Easter meeting. He returned with it to the circuit in October to set up a new 16 hp, flying start, half-mile record at 72 mph. It is just possible to decipher the name *Toodles* painted on the bonnet.

cars, although his second climb in the Open Event was marred by failing petrol pressure and was much slower. More conclusive, as a demonstration of how much quicker *Toodles* had become during the season, his average speed in the final race at Brooklands was 63 mph. He then returned to the track on 6 October to set up a new flying start, half-mile speed record for the 16 hp class at 71.982 mph.

1911: From T-head to L-head and a Directorship

Over the winter the experimental department was busy. In February 1911 a new six-cylinder production car was added to the range available to customers. This new 18-22 hp model had the same bore and stroke as the existing 12-16 hp but with cylinders cast in two blocks of three; it was fitted with an SU carburettor and a worm drive back axle.

In contrast to the relatively conservative production car, Coatalen unveiled his new Brooklands circuit car *Toodles II* in April. The remarkable flying half-mile record of just under 72 mph he had set in October the previous

Wolverhampton Wonders 1909–14

DA 303 was the registration number used on various competition Sunbeams. The original caption describes this car as *Toodles*. Here Coatalen has his wife, Olive, as passenger and Tommy Harrison as riding mechanic as they wait to take part in the Wolverhampton District Automobile Club Hill Climb at Ironbridge in September 1910. He won the Sunbeam and Star Trophies, the RAC medal for best performance irrespective of class and two gold medals on formula.

year on his basically standard 12-16 hp had only stood for a month before it was beaten by G.O. Herbert on a Singer *Bunny Junior* that covered the same distance at 81 mph. Obviously something more than fine-tuning was needed and *Toodles II* was the answer. With this car Coatalen continued his experiments with overhead valves but by limiting the bore of the engine to 80 mm he made sure it remained eligible for the 16 hp class; with its increased stroke length of 160 mm it had a capacity of 3,217 cc. Rather than the four overhead valves per cylinder that *Nautilus* had featured, *Toodles II* had two valves per cylinder but this time they were angled at 100 degrees to provide a hemispherical combustion chamber. To reduce the weight of reciprocating masses to a minimum, the valves were operated by a single, chain-driven overhead camshaft, the conical crowned pistons had cutaway skirts and the connecting rods were extensively drilled. The single-seater bodywork was as narrow as that on *Nautilus* but the nose was formed by a deep v-shaped radiator to provide adequate cooling

Chapter Three

for the powerful new engine. On its first outing at the Brooklands Easter Meeting with Coatalen at the wheel, it won a number of races and then the following day it took back the flying half-mile record at 86.16 mph.

At the Brooklands meeting in May, at the wheel of *Toodles II* Coatalen won two more races and finished third in another. When the Straker-Squire that was following him crashed, it led to suggestions in some quarters that his driving might have been responsible and a protest was lodged. He was fully exonerated by the stewards but his comments published in *The Autocar* give an insight into the state of the track only a few years after its opening. He explained that patches of the track had sunk, which sent the rear wheels shooting up the banking and other raised patches had the opposite effect so that at practically no time did the car pursue a course parallel to the top of the banking. In the previous year the road-holding of one car he drove (possibly *Nautilus*) was so bad that after three laps he said he was completely exhausted,

The engine of *Toodles II* with two inclined valves per cylinder operated by a single, chain-driven overhead camshaft.

having been regularly lifted from his seat by 12 inches until the steering wheel prevented him from going any higher. He had to hang on for dear life, pushing his heels against the special heel board so hard that he burst the back of the seat.[7] The stewards said they hoped to see Coatalen back at the next meeting but he was so upset at being called before them that he declared he was giving up Brooklands for good. It soon turned out that this was a resolution he could not keep for long.

7. *The Autocar*, 3 June 1911.

Opposite: Louis Coatalen poses in his record-breaking *Toodles II* racing car for 1911. It had a chain-driven overhead camshaft and inclined valves.

Chapter Three

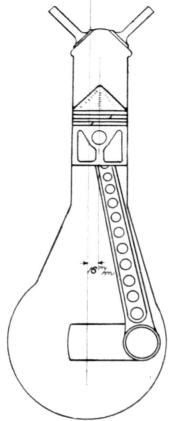

Copyright of] [THE CAR
DIAGRAM SHOWING THE
VALVE POSITION AND
THE DÉSAXÉ-SET

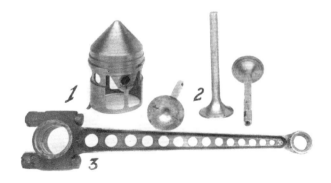

At the beginning of 1911 it had been announced that a Coupe des Voitures Legères race was to be held in France for cars of less than 3 litres capacity. Coatalen saw this as an opportunity to introduce Sunbeams to international racing and in March submitted an entry for one car. By the end of May it was being tested at Brooklands when it became evident that he had effectively produced a new car for the event – the prototype for the revised 12-16 hp Sunbeam that would go into production for the following year. It had a monobloc four-cylinder L-head engine with inclined valves. In his book *The Classic Twin-Cam Engine*, Griff Borgeson suggested that this layout might have been designed for Sunbeam by Ernest Henry or, alternatively, that Coatalen had 'borrowed' some of the design from the Swiss Labor-Picker engine with which Henry had been involved.[8] It seems far more likely that the inspiration came from much closer to home, as Wolseley had introduced a new, monobloc, L-head, inclined valve, 12-16 hp model in October 1909 that was well illustrated and described in the motoring press of the time.[9] Whilst the new Sunbeam shared some details with the Wolseley, such as a single casting for the cylinder block, inclined valves on the near-side and inlet ducts that crossed

Top: Sectional diagram of the engine from *Toodles II* showing the inclination of the overhead valves and the 16 mm offset of the cylinders in relation to the centre of the crankshaft. *Bottom:* The photograph shows a much-lightened connecting rod and piston from the same engine. The weight of the piston, turned from a solid steel forging, with gudgeon pin and rings was 1 lb 7 oz (650 gr).

8. Griffith Borgeson, *The Classic Twin-Cam Engine*, Dalton Watson, 1981, pp. 102-05. The Labor-Picker engine referred to, although it had inclined valves, was a T-head engine unlike the L-head layout of both Wolseley and Sunbeam.
9. *The Motor*, 26 October 1909.

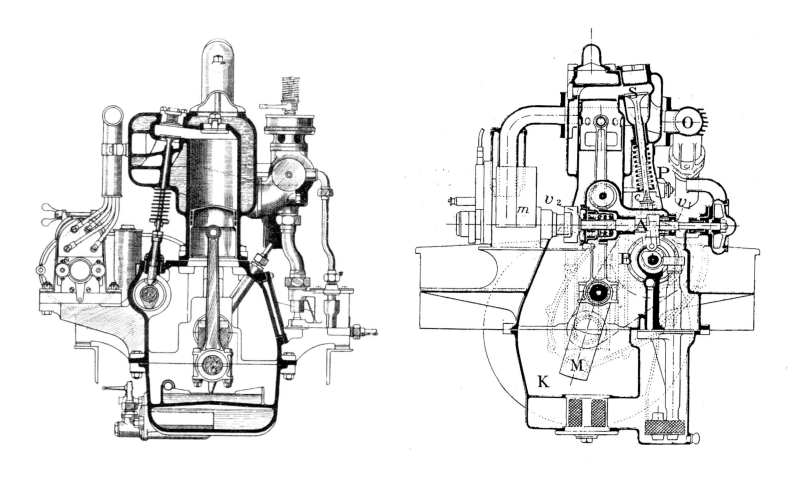

Left: Drawing of the Wolseley-Siddely 12-16 hp engine, published in *The Motor* in October 1909.
Right: Drawing of the Sunbeam 12-16 hp engine, introduced in 1911.

through the block, it incorporated considerable improvements. Whereas the Wolseley had a single inlet duct between cylinders two and three, taking the mixture from the carburettor to the inlet valves, the Sunbeam had two ducts (between 1 and 2 and between 3 and 4); the crankshaft of the Wolseley had only three main bearings and relied on scoops on the con-rods to lubricate big-ends and gudgeon pins – the Sunbeam crankshaft was 'désaxé' running on five bearings and had a gear driven oil pump that pressure fed both main and big-end bearings; cooling on the Wolseley was thermosyphonic – the Sunbeam had a water pump. It is interesting that, at about the same period, Laurence Pomeroy was also experimenting with inclined side valves for the engine that was ultimately developed to power the 30/98 Vauxhall. Most of the other entries for the Coupe de l'Auto race still had cylinders cast in pairs.

Initially it was Coatalen's intention to drive the car himself in the race, held at Boulogne on 25 June, but in the event the wheel was taken by

Chapter Three

T.H. Richards with Coatalen supervising from the pits.[10] (There had been some speculation in the French press that *'la charmante Madame Coatalen'* might not permit her husband to take the wheel.) Approximately two-thirds of the way through the race when it was the leading British entry and battling for seventh place, the car struck a kerb and broke a steering connection, thus prematurely ending Sunbeam's first entry in an international event. However, valuable experience had been gained and the seeds of future ambitions had been sown.

It was not long before other 12-16 hp Sunbeams fitted with the new engine were competing successfully in club events around England but Coatalen's main focus of attention had turned to preparing a major record-breaking attempt in order to publicise the launch of the new 25-30 hp Sunbeam. The 25-30 had a large six-cylinder engine (two blocks of three, 90 x 160 mm, 6,105 cc) which was otherwise similar in design to the new 12-16 hp engine, and for the event was fitted with a single-seater body and baptised *Toodles IV*. On 1 September 1911, at 7.16am, Coatalen started out at Brooklands in beautiful weather to attempt to break the twelve-hour record set up in 1907 by S.F. Edge in a Napier. Sharing the driving

The Sunbeam entered for the 1911 Coupe de l'Auto race at Boulogne with the new L-head engine. Coatalen at the wheel during practice.

---

10. *The Auto-Motor Journal*, 17 June 1911, listing the drivers of all the cars the week before the event wrote, 'Mr L. Coatalen will steer the Sunbeam.'

with T.H. Richards in two-hour shifts they averaged 75.7 mph for half a day and covered 907 miles (Edge had covered just under 800 miles), which was a convincing demonstration of the new car's potential and its reliability. Importantly, this notable event attracted tremendous press coverage with reports of the Sunbeam's achievement appearing in a wide variety of publications at home and abroad in addition to the usual specialist motoring papers. In October Coatalen and the car were back at Brooklands racing against Percy Lambert's Austin, with the Sunbeam winning after a close race. Shortly after that, Coatalen in *Toodles IV* set up a ten-lap standing start '40 rating' record at 84.41 mph and on the same occasion established the '16 rating' record over the same distance with *Toodles II* at 79.29 mph. At this point the flying half-mile record that Coatalen had set in April still stood at 86 mph, so *Toodles II* held both the short and long 16 hp records but this situation did not last long as, in November, both were captured by Tysoe's Singer *Bunny III* and then the bar was set even higher in December by Hancock's special long-stroke Vauxhall at 94 mph.[11]

The revised 12-16 hp Sunbeam production car with its 80 mm bore x 149 mm stroke, L-head engine was submitted to the official RAC Certification Trial and it is interesting to compare the results with those recorded two years before. This time petrol consumption was

Louis and Olive Coatalen at Brooklands on 1 September 1911 for the attempt on the twelve-hour record.

22.65 miles per gallon (1909: 20.12 mpg), even though the 106-mile road test was carried out with the hood up. On the Brooklands circuit the fastest lap recorded was at 65.85 mph (1909: 49.34 mph) one third faster, while acceleration figures of 0-15 mph in 2.82 seconds (1909: 3.6 seconds) and 0-30 mph in 9.10 seconds (1909: 13.96 seconds) showed equally impressive improvements. It was awarded the RAC gold medal for the most meritorious performance in the Monthly Trials of 1911.

Around the time of the Motor Show at Olympia, Coatalen had the opportunity to establish new international contacts. A deputation from the American Society of Automobile Engineers toured Britain and visited the Sunbeam Works among others.

---

11. Nic Portway, *Vauxhall Cars 1903–1918*, records that the target had been to push the record to 100 mph but Hancock was not successful in this endeavour as the special long stroke Vauxhall racing engine fractured its crankshaft.

Chapter Three

During the twelve-hour record, tyres were changed every two hours. Olive Coatalen lends her weight to the improvised quick-lift jack while Louis Coatalen (centre) issues instructions.

It was reported that the English factories such as Daimler, Wolseley, Sunbeam and Humber were judged to be equal to anything in America in terms of equipment and organisation, although they were criticised for the amount of handwork devoted to chassis construction. A deputation of about a dozen motoring personalities also came from France and a light-hearted luncheon was held at Oatlands Park Hotel, Weybridge, on 5 November 1911 with Louis and Olive Coatalen, Henri Perrot and Thomas Pullinger acting as hosts to the French visitors, amongst whom were Charles Faroux, Victor Rigal and Albert Guyot, with the well-known photographer Meurisse in attendance. One of the outcomes of this meeting was that Victor Rigal took on the representation for Sunbeam cars in France and its colonies.

However, the climax of the year for Coatalen came a few weeks later at the Annual General Meeting of the Sunbeam Motor Car Company Ltd, when the chairman, John Marston, reported on the huge amount of progress the company had made. A table groaned under the display of silver cups and medals won by Sunbeam cars during the year and a net profit of £40,997 was announced, which was double the previous figure. The additional capital that had been raised earlier in the year through the issue of new shares had been oversubscribed and the money was used to pay off overdrafts and loans, build a new machine shop and order new machinery and tools. The company had £41,207 cash in hand, compared to just £21 the previous year, output had practically doubled, an additional shop capable of accommodating 400 men had been built, a plating plant and a new

Top: During another wheel change T.H. Richards gets into the driver's seat while Louis Coatale, in the white sweater, goes round the front of the car and Olive Coatalen steps on to the jack lever.
Bottom: Oil level is replenished while the wheels are changed.

'the skipper', who had made the company second to none and praising the decision of the directors to elect Mr Coatalen to the board 'for no cleverer man existed in the motor trade'. In responding, Mr Cureton remarked that the company 'had made wonderful progress in a comparatively few years, and they had to thank Mr Coatalen in particular, for, from the time he joined them they had gone up by leaps and bounds'. Coatalen was never happy as a public speaker, as his English was not that good and it sounds as though he was embarrassed at the attention he was getting because the report says simply 'Mr Coatalen also briefly responded to the toast.'[13]

1912: The Coupe de l'Auto Victory

During the first half of 1912 Coatalen was noticeably absent from the starting grids of races at Brooklands. This was partly a reflection of his new status as a director of the company; the directors were increasingly worried about what would happen to the works should their wonderful Chief Engineer be incapacitated in a racing car accident and exerted much pressure on him. Hugo Massac Buist wrote,

brass casting foundry had been installed, orders in hand were nearly double that of the previous year, and they would have to continue with night shifts. Concluding, Mr Marston said, 'much of the company's success was due to Mr Coatalen, their chief engineer, and they considered it advisable to strengthen the Board of Directors by asking him to take a seat thereon.'[12]

At the first annual staff dinner held at the Victoria Hotel, Wolverhampton, William Iliff (Company Secretary) proposed the toast of 'the directors' paying tribute to Mr Cureton,

---

12. *The Financial Times*, 22 November 1911.
13. Unidentified press cutting, probably a Wolverhampton newspaper (author's collection).

Louis Coatalen at speed on the Brooklands banking, setting up a new twelve-hour record with *Toodles IV* on 1st September 1911. The six-cylinder 6-litre engined Sunbeam 25-30 covered 907 miles at an average speed of 75.7 mph.

At last the Board succeeded in impressing on Mr Coatalen the fact that, though he might design the cars, organise the works, select raw materials, prepare cars for competition, discover talent of all sorts and so on ad infinitum, still somebody else had better drive Sunbeam cars in competitions. He was very reluctant to give up the sport which was particularly dear to him.[14]

However, it seems likely that Coatalen's absence from competition was also partly due to the decision to follow up the previous year's experimental entry into international racing by entering a four-car team for the Coupe de l'Auto race to be run concurrently with the French Grand Prix at Dieppe at the end of June. This was a major undertaking for which preparation was very time-consuming. Coatalen's co-directors may have been reassured that this was not an unreasonable decision by the fact that their ex-employee, Thomas Pullinger, had entered a team of Arrol-Johnstons the year before, as had another British manufacturer, Calthorpe. In 1912 the Sunbeam team would be accompanied across the Channel by teams of Arrol-Johnstons, Calthorpes and Vauxhalls. As early as February, Coatalen and the first Sunbeam were out on the circuit for tests.

---

14. H. Massac Buist, *The History & Development of the Sunbeam Car*, p. 27.

Chapter Three

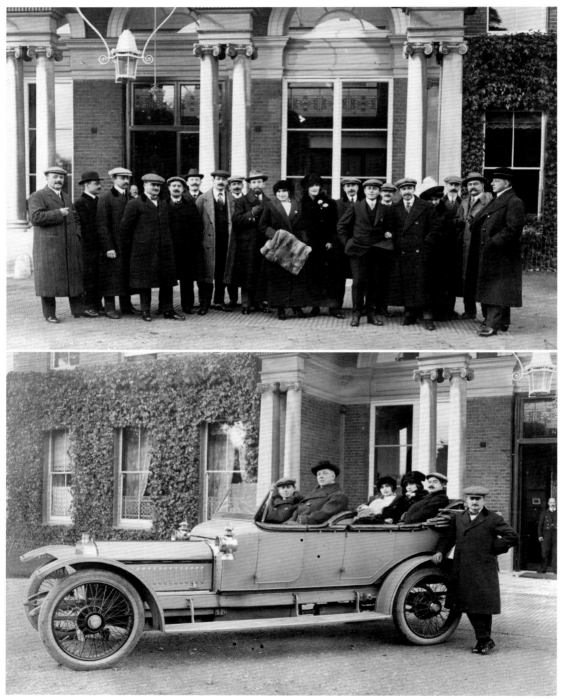

Luncheon for French motorists at Oatlands Park, Weybridge on 5 November 1911.
Top: Charles Faroux is far right in profile, Henri Perrot and Louis Coatalen are arm-in-arm fourth and fifth from right in front row, with Thomas Pullinger behind, and Olive Coatalen in black next to Mme Adda Rigal. Albert Guyot in a bow tie and Victor Rigal are fourth and fifth from left.
Bottom: In the Sunbeam after a good lunch Coatalen, Faroux, Rigal, Mme Rigal and Olive Coatalen with Albert Guyot leaning on the rear mudguard.

Wolverhampton Wonders 1909–14

Left: A menu for the second annual staff dinner celebrating the victories of 1912 has survived. Signatories include Directors; Thomas Cureton, William Iliff, Samuel Bayliss, Edward Deansley, Henry Bath and Louis Coatalen.

With Victor Rigal appointed as Sunbeam's agent in France, it was natural that he was nominated to drive one of the race cars. He was an experienced Grand Prix driver, having taken part in 1907 (Darracq) and 1908 (Bayard-Clément) and no doubt he helped to recruit another old hand, his team mate, Gustave Caillois (1907 Darracq) to complete the team alongside Dario Resta (who drove for Arrol-Johnston in 1911) and the Austrian, Emil Medinger. During that first February trip arrangements were made for the team to be housed during the race close to the circuit in the village of Braquemont. The premises, which belonged to a Parisian doctor, had been used previously by the Renault team in 1907 and this was seen as a good omen. It enabled the whole team, including the cars, to be accommodated together. Coatalen returned in the spring to Dieppe for two further practice periods of a week each when the team experimented with weight distribution on the newly shortened chassis of the racing cars.

The story of the Sunbeam team's momentous historic victory, coming first, second and third in the 3-litre Coupe de l'Auto race and third, fourth and fifth in the Grand Prix itself has been well recorded and need not be detailed here. Coatalen's policy and the decision to enter a team of cars were completely vindicated, as the resulting press coverage was huge and new orders for the production of cars followed.

Left: Louis Coatalen at the wheel of the prototype 1912 Coupe de l'Auto/Grand Prix Sunbeam during testing at Dieppe early in the year. Behind the car are the drivers Victor Rigal, Emil Medinger and Gustave Caillois.

Right: Tea at Les Fresnes, Braquemont, the Sunbeam headquarters for the 1912 Grand Prix. This blurry snap shows that there was time for socialising and relaxing. Left to right: Olive Coatalen, Mme Caillois, Medinger, Mimi Medinger, Mme Aubert, Mme Rigal, Rigal, Coatalen, Searight, Mme Resta.

At the beginning of the second day of the 1912 Coupe de l'Auto Race, Louis Coatalen looks on from the pits as Resta restarts while Rigal's car no. 3 is being re-fuelled.

The achievement in this gruelling race that was two-and-a-half times longer than the previous year's and spread over two days to cover 956 miles, was something for which the whole Sunbeam works could be justifiably proud. In fact, all British motorists were delighted that the team of modest, almost-standard cars had done so well against the pure racing monsters with their chain drive and engines that were more than four times as big. The Royal Automobile Club gave a banquet in the team's honour soon after their return to England, at which the drivers were awarded gold medals and the mechanics bronze ones. In replying to the toast, Louis Coatalen paid great tribute to Brooklands as a facility for developing the cars and stressed how producing a team of racing cars was a good way of crushing out conservatism in a factory and putting everyone on their mettle and interested in the project. The banquet was followed by a 'post-prandial programme that included a cinematograph showing scenes of the race'.[15]

15. *Auto-Motor Journal*, 13 July 1912.

Grandstand pass for the Dieppe Grand Prix.

Back in Wolverhampton the directors met on 3 July 1912 and recorded their delight as follows:

> The Directors at their meeting today wish to give Mr Coatalen, their co-Director, a most hearty vote of thanks for the very skilful manner in which he carried out the necessary organisation in connection with the recent French Race, also to acknowledge their indebtedness to him for designing a car which could achieve such a magnificent result. They congratulate him on the very successful issue and desire that their thanks and appreciation should be recorded in the minutes of this meeting.

Coatalen was given a hand-written copy of the minute, signed by John Marston, Chairman, and they also presented him with a gold repeater hunter pocket watch inscribed with

The victorious Sunbeam team with their enormous bronze trophy. Back row, left to right: Dario Resta, Gustave Caillois, Louis Coatalen, Victor Rigal, Emil Medinger. Middle row, left to right: Eva Resta, unknown, Olive Coatalen, Mme Caillois, Mme Rigal, Mme Medinger. Front row, left to right: Strothers, Smith, unknown, Bate, Harrison, Chassagne.

Chapter Three

his initials on the outer case and inside with the words:

> Presented to Mr. Louis Coatalen by the Directors of the Sunbeam Motor Car Company Limited as a memento of the Victorious Sunbeams in the Grand Prix 1912.

The Sunbeam victory was generally hailed as a great British triumph until a Mr Holloway wrote to *The Autocar*, pointing out that key elements were the result of foreign designers: Coatalen was French, the Claudel carburettor was French, the Bosch magneto was German and all the drivers were foreign too! This provoked an avalanche of heated correspondence about what constituted British manufacture.

A few weeks later Coatalen was letting off steam at Brooklands where he entered *Toodles IV* in the August Bank Holiday meeting. By taking the Byfleet banking high up he overtook all his competitors and won his race by 70 yards. A week after that he took part in the Pateley Bridge Hill Climb at the wheel of one of the Coupe de l'Auto team cars putting up joint fastest time. He then took part on the same car in the Leicester

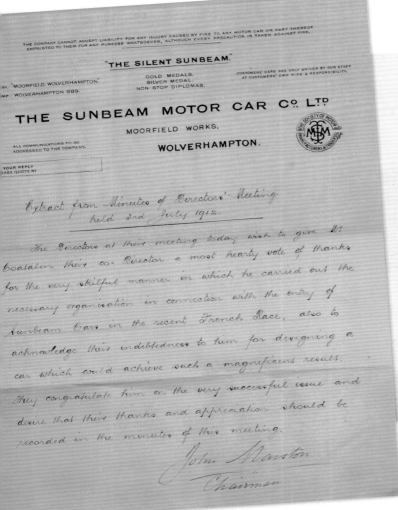

Left: Extract from Minutes of the Meeting of Directors held on 2 July 1912, congratulating Coatalen on his success at Dieppe.

Right: Menu cover for the banquet organised by the Royal Automobile Club to celebrate the Sunbeam victory.

Wolverhampton Wonders 1909–14

Left: Coatalen (right) looks on as Resta is pushed off before a record-breaking run at Brooklands in September 1912. One of the 3-litre Coupe de l'Auto cars had been fitted with a narrow streamlined single-seater body and disc wheels. On the second attempt the World Record for 1,000 miles was set in 13 hours and 8 minutes at an average of 76.10 mph and the twelve- and thirteen-hour records were also broken. The fastest lap was covered at 86.77 mph.

Right: Louis Coatalen and Searight (of Claudel Hobson) put the bonnet on the single-seater 1912 Coupe de l'Auto car before the record attempt. Dario Resta leans over the cockpit.

A.C. Hill Climb at Beacon Hill, Woodhouse Eaves, with his wife Olive in the passenger seat. This seems to have been the extent of his own competitive motoring for the year, apart from taking part in one record attempt.

In September and October successful attempts were made on records at Brooklands with both *Toodles IV* and one of the Coupe de l'Auto cars fitted with a streamlined single-seater body but most of the driving was delegated to Dario Resta and R.F. Crossman, although Coatalen shared the wheel during the first attempt. With the smaller car they established a twelve-hour record and later finally pushed the 16 RAC Rating record for the flying half-mile to over 100 mph. In two years the Sunbeam speed over that distance had increased by 30 mph or some 40 per cent.

At the Annual General Meeting in November Mr Marston was able to report to shareholders that the company was producing between thirty-five and forty cars per week – a car taking about six weeks to complete – and they had nearly 2,250 employees. The machine shop had been doubled in size, a new body shop and a new paint shop had been built while further improved facilities were in hand. He was reported as saying that the increase in business was due to the splendid achievements of their cars in the Grand Prix and Coupe de l'Auto races for which 'Mr Coatalen was absolutely answerable' (even though the financial year

Chapter Three

ended in August and so the race result cannot have had much effect). The record net profit for the year of £60,889 was down to the skilful management of Mr Cureton, Managing Director, and enabled the company to declare a 25 per cent dividend.

Around this time the Chief Engineer at Vauxhall, Laurence H. Pomeroy, wrote an article in *The Autocar* arguing that a 3-litre engine with 90 x 120 bore and stroke was superior to an 80 x 150 3-litre engine. He had just redesigned the engine of the A-Type Vauxhall and obviously felt the need to explain why he had retained the 90 x 120 dimensions of this engine despite the fact that it pushed the car into a higher taxation bracket (50 per cent more than the Sunbeam) which was still based at that time on the cylinder bore size and not overall cubic capacity. Pomeroy even suggested that a 'square' engine of 98 x 98 would be ideal but in order to provide sufficient bearing surfaces the engine would have had to be unduly long and the short stroke would have resulted in utilising such short valve springs that these might be unreliable. He stated that the 90 x 120 was a happy mean. In contrast, he argued that an 80 x 150 engine was inevitably 'more gawky' in design with longer crank webs, a wider and deeper crankcase, longer connecting rods and a heavier flywheel. He was convinced that a stroke to bore ratio of 1.33:1 was superior to a ratio of 1.89:1 for no appreciable difference in maximum horsepower.

As the successful 12-16 hp Sunbeam had an 80 x 150 engine, Coatalen immediately responded to defend his choice of this ratio of stroke to bore. He wrote that for the Coupe de l'Auto race,

> I had the option of adopting any ratio from the square engine, in which bore equalled stroke, to that in which stroke was double the bore – after careful experimenting I deliberately chose what was practically the biggest stroke to bore ratio allowed by the regulations.

He argued, perhaps with an amount of poetic licence, that with the smaller bore 'power is delivered more softly and is of longer duration' and is 'consequently nearer to steam power in action'. This enabled the weight of the moving parts to be reduced and he even went so far as to say, 'the weight of the piston is such an important item that I prefer to save one ounce in the piston than 1 cwt in the car'. He pointed out that his design had been severely tested and proven. In the race Sunbeam was the only team which had three cars at the finish despite being kept running between 2,400 and 3,200 rpm for 900 miles. 'The 80 x 150 engine has shown itself to be a far better stayer … durability is what motorists want.' The sub-text, which was not mentioned but would have been only too clear to the two men, was the fact that the whole Vauxhall team had retired from the race as a result of failures in their 90 x 120, 3-litre engines.

However, Pomeroy was unconvinced. He replied that Coatalen's arguments were based upon racing experience and nothing else. 'I have found no serious repudiation of my claim that the 90 x 120 is better than the 80 x 150 for touring purposes,' and suggested that Coatalen's arguments were largely based on fallacious premises.

Louis Coatalen had concluded his article with the statement, 'we all remember how severely criticised was the advent of the magneto, the ball bearing, the pressed steel frame and forced lubrication, all of which are universally used today. I venture to predict the same result for the long stroke engine which marks a real forward step in motor construction.' A hundred years later it is evident that he was wrong and Pomeroy was right but at the time the Sunbeam was extraordinarily successful and there was then little reason to think otherwise.[16]

Another significant event in 1912 was Coatalen taking out his first-ever patent in conjunction with the Sunbeam Motor Car Co., entitled 'Improvements in Starting Systems for Internal Combustion Engines'. It was for a compressed air starter which had a three-cylinder engine designed with 120 degree crank throws so it had no dead centre. Although it was offered on cars, it became even more useful on the big aero-engines.

Coatalen was in Paris just before Christmas 1912 for the Motor Show and so was able to attend the fifth annual banquet of the *Gadzarts de l'Automobile* (graduates of the Arts et Métiers schools working in the automobile industry) held at the Taverne de Paris, place Clichy, on 21 December. He must have been a little surprised to find himself one of the heroes of the evening, having graduated from the school under a cloud, but a toast was drunk to those alumni who had distinguished themselves in motor racing, namely Coatalen, Delage and Brasier. After speeches by Mr Sainturat of *La Vie Automobile* and Mr Corre, Director of the Paris Arts et Métiers School, Coatalen was invited to address the seventy colleagues present. The report of the proceedings notes that he was somewhat shy and having spent so long in Wales (sic) had lost the habit of talking French. He was described as clean-shaven, having a determined chin, and a profile from a roman medallion. He paid tribute to Brasier and Delage, whose example he had followed and the report goes on to describe how, under his energetic impetus, production at Humber had risen from fifty cars a year to 2,000 a year in six years. Since joining Sunbeam, production had risen from 100 to 2,500 a year, in four years. The conclusion was that such feats by one *Gadzart* reflected glory on them all and served as a stimulant to the young and a comfort to the old.[17]

1913: Boats and Planes

After a glorious year in 1912, Coatalen entered 1913 full of plans to carry on to even greater heights in motor racing but also to drive the Sunbeam works forward by expanding its scope through manufacturing marine and aero-engines. Naturally enough, these three strands were intertwined; boat and plane engines appeared on land and vice-versa.

Even before the end of 1912 the first prototype V8 aero-engine had been running on the test bed in the works. It was based

---

16. *The Autocar*, 16 November 1912, 7 December 1912, 14 December 1912.
17. Y. Guédon, *Cinquieme Banquet des Gadzarts de l'Automobile et des Industries qui s'y rattachent*, 21 December 1912.

Chapter Three

on the design of 12-16 hp car engine (80 x 150 mm, L-head) but with cylinders cast in pairs, mounted at 90 degrees on a common crankcase, and with the slightly inclined valves operated by a single central camshaft. For marine use it was modified to have two blocks of four cylinders with the exhaust pipes rising vertically, unsilenced, from the centre of the V. In March 1913 a pair of these marine engines was shipped out to the boatyard of Alphonse Tellier on the Seine to be installed in a hull he was constructing for Mr Philippe Leo. The boat was a hydroplane called *Sunbeam* and it was to be entered in the annual week of motor boat races to be held at Monaco in April as a way of testing them under load.

At the opposite end of the spectrum a special engine was developed for the British Motor Boat Club's new 21 ft class, at the same event, which limited engine size to 2½ litres with a maximum stroke of 6 in. For this class Coatalen provided a short stroke (120 mm) version of the successful 12-16 hp four-cylinder engine and at least five of these engines were supplied as there were that number of boats in the 21 ft class with Sunbeam engines that year. This was in an entry of eighteen boats of which three each were powered by Brooke and by Austin engines, there were two with Wolseley engines and one lone Vauxhall-driven boat. The remainder had French or Italian engines.

Top: The first V8 Sunbeam marine engine for the hydroplane *Sunbeam*. The slight inclination of the valve springs is evident in this picture.

Bottom: *Sunbeam* tied up to a French Naval vessel at Monaco in April 1913. On board are Messrs Tellier, Dubonnet, Coatalen (standing), Chassagne (seated), and the owner Mr Leo peering into the engines.

Wolverhampton Wonders 1909–14

The smaller Sunbeam-powered boats for the 21 ft class in line astern at Monaco.

Although there is nothing to hint that Coatalen had ever been to Monaco before, he seems to have been instantly at ease socialising with his French contemporaries and taking the helm of boats, which was a new sport for him. He was accompanied by his wife Olive, his boss, Alderman John Marston JP, and backed up on the mechanical side by Jean Chassagne. Olive, who by then was five months pregnant, found herself in the company of the wives of well-known Frenchmen, such as Mesdames Bleriot, Delage, Dubonnet, Janvier and Tellier.

The twin-engined *Sunbeam* showed potential but ignition problems were experienced and she could only manage a third place in the Prix de Monaco race and retired early from the Coupe des Nations. Fortunately, the Sunbeam-powered boats in the 21 ft class put up a more impressive display, achieving three wins, one second place, two third places and two fourth places. Coatalen himself was at the wheel of Ernest Martin's *Fujiyama III* when it won the Prix du Premier Pas over 50 km. He was also photographed driving Oscar Martin's *Vicuna III*, which started in the Prix de la Méditerannée and the Grand Criterium des 21 Pieds but it was either not placed or retired in both these events (it had run a big end bearing in the Premier Pas race).

The development of the aero-engines will be pursued in more detail in the next chapter but during the programme of test and development one of the V8 engines was shoehorned into a standard Sunbeam chassis. At the Shelsley Walsh Hill Climb held in June, both C.A. Bird and Louis Coatalen drove this potent special, putting up times of 58.4 seconds and 60.2 seconds respectively but this was a one-off appearance. More significant was the

Louis Coatalen on *Fujiyama III*.

Top: Shelsley Walsh Hill Climb, 7 June 1913. Louis Coatalen at the start on the experimental car fitted with a V8 engine. His fastest time was 60.2 seconds. C.A. Bird driving the same car climbed in 58.4 seconds.

Bottom: Jean Chassagne at the wheel of the V12, 9-litre engine Sunbeam *Toodles V*, having just set the World Record for one hour at 107.95 mph at Brooklands. Louis and Olive Coatalen stand on the left; next to Chassagne is thought to be Mr Thisleton of Dunlop. The car was later raced in America.

development of a 9-litre V12 aero-engine which was fitted into a 25-30 hp chassis for record-breaking purposes by the end of August. Named *Toodles V* and fitted with a slim single-seater body, it first took part in races at Brooklands with Jean Chassagne at the wheel in October. He won the Long Handicap at an average of 110 mph which was the fastest-winning speed recorded up to that point and put up a fastest lap at over 118 mph in the Short Handicap, finishing third. A few days later Chassagne was back at the circuit to set up records for 50, 100 and 150 miles and to take the one-hour world record at 107.95 mph. Around the same time one of the V8 engines was fitted into a Henry Farman biplane and took its first flights.

In parallel with these novel nautical and aeronautical developments Coatalen embarked in 1913 on the most ambitious motor racing programme yet to be undertaken by the firm. That year the Grand Prix de l'ACF and the Coupe de l'Auto reverted to being run separately on different circuits and it was decided to enter teams in both races. In view of this it was wise not to be also tempted into entering a team for the Indianapolis 500 Mile Race. Charles Sedwick had crossed the Atlantic in November 1912 with the sole objective of obtaining entries and he tried to tempt European manufacturers to send cars across to the USA by underlining the amounts of money that could be won. It was announced that 'after careful consideration' the Sunbeam directors had decided not to take part but a few weeks later it emerged that Louis Coatalen had agreed to lend *Toodles IV* to Albert Guyot who was to drive it, rebuilt with a two-seater body, as a private entrant. This he duly did and finished fourth, winning $3,500. With its long chassis it was not ideally suited to the Indianapolis bowl and it was outpaced by Goux's winning Peugeot, which had the additional advantage of a somewhat larger capacity engine (7.4 litre compared with the Sunbeam's 6.1 litre).

By the end of April Coatalen's new 4.5-litre six-cylinder Grand Prix car was well advanced, and on his way back from Monaco he met up with his drivers to carry out fuel consumption tests on the Grand Prix circuit at Amiens. Once back in England he then took part with one of the Grand Prix cars on 3 May at the Lancashire Rivington Pike Hill Climb in pouring rain, and a couple of weeks later he and Resta were out with it again for tests at Brooklands. This coincided with Coatalen setting up an unofficial Class C, ten-lap record at the wheel of the single-seater 1912 Coupe de l'Auto race car fitted with a short stroke (120 mm) engine running experimentally on benzole produced from coal tar.

The design of the six-cylinder, side-valve L-head, engine for the Grand Prix de France to be held in July was essentially an extended, improved, version of the previous year's Coupe de l'Auto winning cars in a longer chassis fitted with a differential-less axle. Once again accommodation for the whole team during the Grand Prix was sought and this year Caillois found and rented a semi-derelict château at Moreuil, complete with dungeons and a twelfth-century chapel. This was far less convenient than the house and gardens at Braquemont the previous year that had proper garages and a good workshop. Not only was it reputed to be haunted but when they arrived there was not a stick of furniture or vestige of crockery in the château, and the cars had to be worked on outside under the trees. The team made themselves at home as best they could and soon it became a sort of social centre for people involved with the race. When a visitor arrived, 'Le Patron', Louis Coatalen, 'would

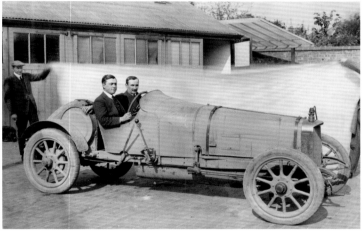

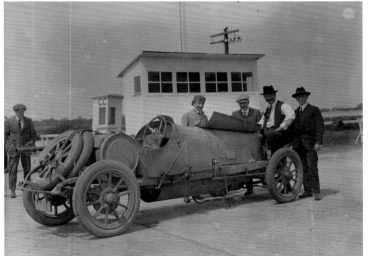

Top: Louis Coatalen and Jean Chassagne in one of the newly finished 4½-litre, six-cylinder Sunbeams for the 1913 French Grand Prix.

Centre: A French group tuning up a 1913 GP Sunbeam at Brooklands. Left to right: Jean Chassagne, Henri Perrot, Henri Claudel and Louis Coatalen.

Bottom: During practice for the 1913 Grand Prix at Amiens, Louis Coatalen is at the wheel of the car that Kenelm Lee Guinness drove in the race with his mechanic Cook.

Chapter Three

Left: Coatalen (centre) in discussion with Dario Resta (left) and Gustave Caillois (right).
Right: Louis and Olive Coatalen with their six-cylinder Sunbeam coupé at the Château de Moreuil.

seize a telescopic-sight gun and sallying forth would return in a few minutes with a rabbit'. After dinner in the big dining hall with its dark oak ceiling bedecked with fleur-de-lys dimly illuminated by candles, Olive Coatalen would put on an 'allegedly humorous' ten-minute bioscope show that she had discovered.[18] Algy Lee Guinness probably had more comfortable lodgings, as he had sailed his yacht *Perlona* up the Somme Canal as far as Amiens to support his brother Kenelm, driving for the Sunbeam team for the first time. The Grand Prix de l'Automobile Club de France was run in July under a fuel consumption limit of 14 mpg, which excluded the 'monster-engined' cars of previous years but seems to have had little impact on limiting the speed of those cars taking part. The winner was Boillot, who averaged 72 mph in a Peugeot, followed in second place by his team mate Goux, with Chassagne in the first Sunbeam coming in third, just four minutes behind after eight hours' racing. The results of the other Sunbeam team members were less brilliant: Resta finished sixth behind the two Delages, Caillois retired and Guinness had an accident when a tyre burst on a sharp corner.

In September the Coupe de l'Auto race for 3-litre cars was held at Boulogne and Sunbeam entered a team of three cars for Chassagne, Resta and Guinness. It was a considerably shorter race (387 miles compared to 956 miles

18. *The Autocar*, 19 July 1913.

Olive Coatalen models her stylish coat and hat.

Wolverhampton Wonders 1909–14

Grandstand badge for the 1913 Grand Prix de l'ACF at Amiens.

in 1912), so there were no qualms about the slightly improved cars standing the test, and the team arrived full of optimism. The engines had been lightened, so they now produced 87 bhp (up from 74 bhp) and had a top speed of 95 mph (up from 85 mph). The only other significant modification when compared to the previous year's victorious cars was the use of differential-less rear axles but this turned out to be their weakness as both Resta and Chassagne were forced to retire with broken rear axles. Only Guinness, who had recovered from his accident at Amiens, kept going throughout the race and finished in third place behind Boillot and Goux on their Peugeots and ahead of Hancock's Vauxhall.

Louis Coatalen went back to France for the Paris Motor Show, which in 1913 opened on 17 October. The centrepiece of the Sunbeam stand was the six-cylinder record-breaker, which had recently covered over 1,000 miles in twelve hours. Coatalen again attended the Gadzarts' dinner at the Café de Zimmer-Madrid with his friend and fellow Gadzart, Henri Perrot. This time it was Louis Delage who was called to speak, as his cars had won the Grand Prix de France (organised by the A.C.O. at Le Mans) against a team of Mercedes-Benz. Delage paid tribute in his speech to 'notre camarade Coatalen' for, although Sunbeams had not won the other Grand Prix (organised by the A.C.F.) at Amiens they had beaten the Delages. In his comments on Coatalen he drew attention to a whole series of records just set up by Chassagne at Brooklands with the V12-engined *Toodles V* and the six-cylinder single-seater, so Delage was not unduly flattering in saying that Coatalen 'is considered as one of the best automobile engineers in the whole of England'.[19] Naturally there was some pride in the fact that he was French and one of their alumni and it is interesting that although Peugeot had won the A.C.F. Grand Prix and the Coupe de l'Auto Race, they were never mentioned, as none of their engineers were Gadzarts!

Following the great racing success in 1912 and the publicity this had generated for the marque, business for the Sunbeam Motor

Dinner at the Château de Moreuil, 20 June 1913.
Standing, left to right: Caillois, unknown, Chassagne, Maeght.
Seated, left to right: Guinness, Olive Coatalen, Louis Coatalen, Mrs Resta, Dario Resta.

---

19. Yves Guédon, *Sixième Banquet des Gadzarts de l'Automobile etc.* 18 October 1913.

Chapter Three

Top: Coatalen filling the petrol tank of K.L. Guinness's car for the 1913 Grand Prix. Cook, the mechanic, with back to camera, holds the funnel while Guinness checks the radiator. In the race a tyre burst and the car crashed, killing a spectator. Guinness and Cook were not seriously injured despite plunging into the river.

Bottom: Message from Jean Chassagne to Mr and Mrs Coatalen in recognition of the excellent car he had been entrusted with for the 1913 Grand Prix at Amiens and regretting that he had not been able to win. He finished third.

Car Company had been excellent in 1913. At the Annual General Meeting a further record-breaking net profit of £95,000 was announced, compared with just £90 profit five years previously when Coatalen had joined the company. An unprecedented 33⅓ per cent dividend for the shareholders was declared. It was not felt necessary to introduce any new models for 1914, although a few minor improvements were made to the existing models.

1914: Twin Overhead Camshafts

The dominance of Peugeot throughout 1913 undoubtedly rankled with Coatalen. Peugeots had been faster than Sunbeams at Indianapolis and at the Grand Prix but both of these could be explained away as being due to bigger engines. However, at the Coupe de l'Auto where everyone was under a 3-litre limit, this excuse no longer washed. And finally, just to rub salt into the wound, Jules Goux had brought one of the 3-litre Peugeots onto Coatalen's home turf at Brooklands to take the flying start half-mile record at 106 mph (Sunbeam in 1912 had managed 102 mph) and push the standing start ten-lap record up to 100 mph (Sunbeam's speed in 1912 was 94 mph). In addition to being impressed by their performance, Coatalen was probably equally envious of the conditions under which the cars were built. The Peugeots had been made in a works set up specifically to produce the ultimate racing machines, far from the production car factory, under the control of

A line of 12-16 Sunbeams ordered by the Russian Government awaiting despatch to Archangel in 1914. Kenelm Lee Guinness and Louis Coatalen can be seen behind the cars.

racing drivers Georges Boillot, Jules Goux and Paul Zuccarelli, with Ernest Henry as their 'brilliant young engineer' and a budget of £50,000. It was the beginning of motor-racing professionalism. Equipped with twin overhead camshafts and four valves per cylinder these were the ultimate racing engines of the day and they had enormous influence on engine design for years after.

After the 1913 Coupe de l'Auto race Laurence Pomeroy, the Vauxhall engineer, said to Louis Coatalen with a smile, 'I think I see some of us redesigning our engine tops for competition work. It seems as though we had better look up some of our old drawings.'[20] Coatalen must have been conscious of the limitations of his own overhead valve engine in *Toodles II*, which had been rapid when first raced almost three years before but was now no match for the increased speed of the Peugeots and, anyway, it had never been tested in anything other than short races at Brooklands.

---

20. Massac Buist, *The Autocar*, 27 September 1913, p. 566.

Chapter Three

It is not surprising that he wanted to discover the internal secrets of the Peugeot engine to see what made it go so well, so he set about acquiring one to study.

The story of Coatalen 'borrowing' and copying one of the Peugeots has developed into such a romantic 'urban myth' that it is opportune to re-examine the evidence and what we know. The seeds of the myth were sown long ago but the story has become exaggerated to such an extent that I have heard it said that Dario Resta parked a Peugeot by the roadside in Wolverhampton one afternoon and came back to collect it the next morning. Overnight it was supposedly taken to Louis Coatalen's home, dismantled, measured and drawn up, then reassembled and driven back to where it had been found and no one noticed its absence. This is simply not true and would be impossible to achieve in such a short time.

It is known that Peugeot disposed of their 3-litre Coupe de l'Auto racing cars at the end of the 1913 season. It had been announced that the capacity limit was to be reduced to 2.5 litres, so the cars were obsolete. Jacques Menier of the French chocolate-making family bought one, which was raced by Duray at Indianapolis in 1914. Two others were raced by private owners at Brooklands. No less an engineer than Henry Royce was as curious as Coatalen to know what made them tick and bought one to study in the autumn of 1913 (he retained it until the beginning of the war) so it would not have been too difficult for Coatalen to acquire one as well.[21] In addition to the contacts of Dario Resta and Jean Chassagne, it should be remembered that Victor Rigal, Sunbeam's agent in France, had driven one of the Peugeots in the race at Boulogne. It seems probable that Coatalen's name and that of the Sunbeam Motor Car Co. would have been kept out of the transaction, so officially it was bought by someone else, we know not who, and so it was said to have been 'borrowed'.

That Coatalen would want to keep quiet the fact that he was studying his competitor's design is understandable. Anthony Heal, who researched the subject in depth, recorded that it 'was still discussed by Sunbeam ex-employees only in hushed tones even half a century later' and when Bill Boddy published the story in *Motor Sport* in November 1977 he could claim it as a scoop. When writing to Anthony Heal in 1943, some thirty years after the events, Hugh Rose had said simply, 'We obtained a Peugeot by some means and took it to pieces (I won't say any more and this is not for publication please).'[22] A letter from Frank Bill, one of Sunbeam's skilled racing mechanics, confirmed that the Peugeot was taken to Coatalen's private home, Waverley House, in late 1913 and that the operation took a long time: 'two of us and a draughtsman spent *many secret months there* (author's italics) and removed every item in detail'.[23] Furthermore, a letter from Laurence Pomeroy junior, technical editor of *The Motor*,

---

21. In addition to the Peugeot, Rolls-Royce also purchased, via their French subsidiary in July 1914, one of the 1913 Mercedes racing cars powered by a modified six-cylinder aero-engine. He subsequently was able to examine in detail one of the 1914 Mercedes Grand Prix cars. See Derek Taulbut, *Eagle*, RRHT, 2011, pp. 14, 279.
22. Hugh Rose letter to Anthony S. Heal, 1 June 1943 (author's collection).
23. Frank Bill letter to Anthony S. Heal, 7 February 1943 (author's collection).

tends to confirm that the Peugeot had been purchased and not just borrowed. He wrote: 'L. Coatalen is supposed to have had one of the 1913 Peugeots purchased ... he then had a draughtsman make a copy of it down to the smallest details. How far this process was taken is shown by the fact that the Peugeot was used by Sunbeams during the last war as a staff car and after many thousands of miles seized its camshaft. It was possible to repair this very easily by removing the camshaft housing from one of the T.T. Sunbeams and putting it on the Peugeot.'[24] Percy Mitchell, the other witness to the events, interviewed by Boddy and Heal in 1977, who worked on dismantling and reassembling the Peugeot with Frank Bill, told the story that Dario Resta had been given the task of driving the car round to dealers' showrooms as a publicity exercise and then arranged for Chassagne to drive it to Wolverhampton. Mitchell said that once it had been rebuilt it was taken to the works to be tested and the implication was that this was after some months rather than days.

With not much time left before the beginning of the season, Coatalen was faced with the dilemma of making a new design incorporating some of the information he now had at his disposal from the Peugeot or alternatively producing a scaled-up copy of the Peugeot engine. He chose the latter path, remarking with some justification to his

Waverley House, the Coatalen's home in 1913 where the Peugeot was dismantled and examined.

colleague, 'Pomeroy, ee ees a vise man who copy wizout altair.'[25]

What was perhaps less wise was the very ambitious programme of racing and development he undertook that year. In addition to building a team of 3.3-litre cars to take part in the Tourist Trophy race on the Isle of Man in mid-June and another team of 4.5-litre cars for the French Grand Prix at Lyons in early July, a short-stroke version was made for boat racing in May. Simultaneously the aero-engines were also still under development and being tested in flight throughout the early part of the year. The two Sunbeam aero-engines, the 150 hp V8 and the 225 hp V12 were on display at the Aero Show at Olympia in March at the same time as the directors announced the ordinary share capital of the firm was to be doubled. To underline the abilities of the aero-engines, Chassagne driving *Toodles V*

---

24. Laurence Pomeroy letter to Anthony S. Heal, 21 March 1940 (author's collection). This is further confirmed by a letter from A.P. Geddes, a nephew of W.M. Iliff, who wrote to Anthony Heal in 1950: 'When I was rummaging round the works in 1919, I came across a Coupe de l'Auto Peugeot in a corner of the repair shop under a lot of junk.'
25. Laurence Pomeroy, *The Evolution of the Racing Car*, William Kimber, 1966, p. 82.

Chapter Three

Top: Coatalen at the helm of *Toto* at Monaco in April 1914, accompanied by Jean Chassagne. The boat was fitted with a short stroke, 16-valve, twin-overhead camshaft engine.

Bottom: Coatalen, A.J. Holder (owner of *Toto*) and F. Gordon Pratt (designer of the hull) at Monaco.

powered by the 9-litre V12 engine, set up new class records for the half-mile and the mile at 119.76 mph and 120.73 mph respectively.

By April the new 'Peugeot-inspired' 16-valve four-cylinder engines were being bench tested and the 2½-litre short-stroke version was fitted into a racing boat *Toto* for its owner Mr A.J. Holder. Louis and Olive Coatalen, accompanied by Jean Chassagne, travelled to Monaco to take part in the annual motor boat races there. One report suggested that 'neither hull nor machine had left the drawing office six weeks before the meeting' so this was the first opportunity to test the engine under stress. Coatalen himself took the helm in a number of races and *Toto* won two, achieved a second place and a fourth place in other races. The only problems they encountered were with the hull in rough seas but the performance of the engine was very encouraging. There were three other Sunbeam powered boats (*Youki*, *Vicuna IV* and *Frigidi-Pedibus*) at the meeting, which also performed creditably but these were equipped with the side-valve engines.

On their way back from Monaco, Coatalen and Chassagne stopped off at Lyon to reconnoitre the Grand Prix circuit where they were joined by Kenelm Lee Guinness and Dario Resta. The circuit had already been closed for testing by any racing cars so they could only tour around to assess its challenges. Chassagne was able to stay for a week, getting to know it with a touring car before taking a boat to the USA where he was to compete at Indianapolis, while the others rushed back to England to fine-tune the cars for the Tourist Trophy race. An article on engine design written by Louis Coatalen was published in *The Motor* on 19 May. Perhaps this was an attempt at obfuscation to distract the general public from the source of the design for his new racing engine, but it explained how Sunbeam race engine valve gear

Wolverhampton Wonders 1909–14

Left: Wearing different hats, Coatalen and Pratt pose with Olive Coatalen alongside *Toto*.
Right: Louis and Olive Coatalen relaxing on the terrace of the Beau Rivage Restaurant, Monaco, with Gordon Pratt in April 1914.

had evolved. Although it had been known for some years that more power could be obtained by using overhead valves, Coatalen claimed that for long-distance races he had sacrificed power for reliability by using side-valve engines. Now with improvements in actuating the valves it was possible to benefit from the advantages that overhead valves offered; more power due to less heat loss, improved combustion chamber shape and by using four valves per cylinder greater valve area and therefore more gas flow. He recalled how in 1910 *Nautilus* had had the first Sunbeam overhead valve engine employing four valves per cylinder but it had required 'abnormally strong valve springs'. In 1911 *Toodles II* had followed with an overhead camshaft driven by chain that had proved unsatisfactory. He then said that in 1912 another engine was designed with the camshaft driven by a train of straight spur wheels, which was believed to be the first of its kind.[26] For 1914 the camshafts were mounted directly over the valves. Coatalen pointed out that Maudsley had been using shaft-driven overhead camshafts since 1903 but somewhat tongue in cheek he also mentioned 'in passing it should be noted that a similar type of engine was adopted by the Peugeot Co. in their very successful racing cars'.

Ernest Henry had designed for the Peugeot twin overhead camshaft, four-valve-per-cylinder racing engines, an unusual L-shaped cam follower, which Coatalen and his craftsmen duly copied. However, Coatalen soon saw that it could be improved upon by replacing it with a hinged finger-type cam follower not unlike that which he had already used on the original

---

26. No other details of this engine are known and there is no record of it having been raced. However, it is noteworthy that in the series of cars named *Toodles* we know of *Toodles*, *Toodles II*, *Toodles IV* and *Toodles V*. Perhaps it was *Toodles III*.

Chapter Three

One of the new Sunbeams built for the 1914 Isle of Man Tourist Trophy Race with Louis Coatalen and Works Manager C.B. Kay.

T-head 12/16 Sunbeam.[27] For the article in *The Motor*, a diagram showing these finger-type cam followers was used even though only one of the cars for the Tourist Trophy race incorporated this feature. Subsequently the engines for the Grand Prix were all fitted with them.

Early in June the Sunbeam team installed themselves at the Fort Anne Hotel, Douglas on the Isle of Man in preparation for the Tourist Trophy Race, which was being run again for the first time in six years. With both the Guinness brothers and Dario Resta as drivers, as well as Coatalen's personal experience of the course, they had the advantage of plenty of local knowledge. Practice was only allowed from 4.30am to 7.00am, so there was also time for socialising and playing golf during the day despite plenty of rain and strong winds. As usual, Coatalen was accompanied by his wife Olive but this time in addition there was 'a chubby faced little fellow', their eight-month-old son Hervé, who was delighted by the noise of the engines. It is on record that Louis took Olive out for a lap in the spare car, but presumably not the baby.

Even if international interest in the Tourist Trophy race was limited, it was nonetheless a very challenging and gruelling race intended to show that Britain too could mount a Grande Epreuve. For a total of ten and a half hours over two days racing, the cars had to climb 1,500 feet to the top of Snaefell and down again on the sinuous island roads to complete the 37½-mile lap sixteen times. An indication of the difficulty of the course can be gathered from the fact that the winner, Kenelm Lee Guinness, covered the 600 miles at an average speed of 56 mph, which can be compared to Rigal's 65 mph average in another long distance race, the Coupe de l'Auto at Dieppe, two years before. Resta's Sunbeam, which incorporated some modifications to its engine when compared to the original Peugeot design, went out early in the race but the other two dominated in first and second places until Algernon Lee Guinness had to retire with a seized universal joint just three laps from the end. However, with Kenelm Lee Guinness finishing first with his Sunbeam this was an encouraging debut for the new engines and the team must have been fairly optimistic as they rushed from the Isle of Man back to Wolverhampton to collect the Grand Prix cars and set off to Lyon.

The Sunbeam Board of Directors were once again delighted with the victory and when

---

27. See drawing in Dowell and Richens, *Sunbeam the brass period*, p. 175.

Top: The Sunbeam team outside the Fort Anne Hotel, Douglas. Left to right: L. Hornsted and C.B. Kay (in the spare car), Mrs Resta, Capt. Toby Rawlinson, Louis Coatalen, H.W. Bunbury, K.L. Guinness and J. Cook (in car 4), Thornton Rutter (Daily Telegraph), Olive Coatalen, Dario Resta and Tom Harrison (in car 15), A.L. Guinness and Smith (in car 21).

Bottom: The team of Sunbeam mechanics in the yard behind the Fort Anne Hotel. Left to right: unknown, Tom Harrison, J. Cook, C.B. Kay, unknown, B. Bate, Smith, Murray (?)

Chapter Three

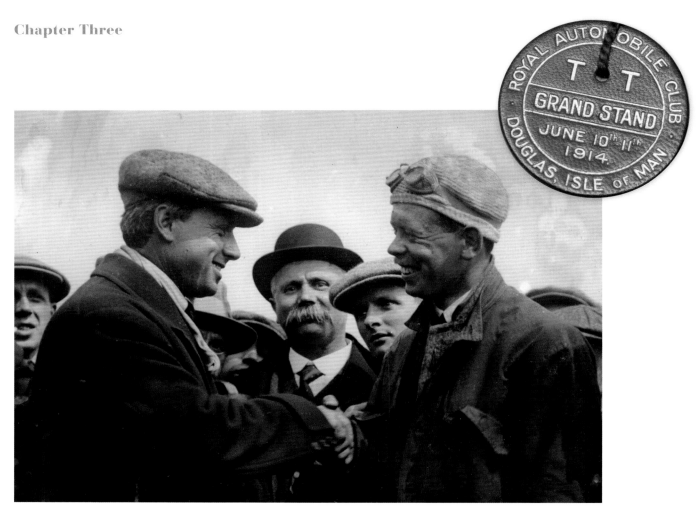

Left: Coatalen congratulates Kenelm Lee Guinness on winning the 1914 Tourist Trophy Race.
Inset: Grandstand pass for the TT Race.

they met on 16 June in the week after the race, they passed the following resolution:

> RESOLVED that the Directors have pleasure in placing on record their appreciation of Mr Coatalen's excellent design of the Racing Cars used in the recent Tourist Trophy Race and the admirable manner in which he conducted the arrangements and desire to accord him their unanimous thanks and their congratulations on the result of the race.

> They also desire to place on record their appreciation of the skill of the drivers and thanks for the praiseworthy efforts of each of them to uphold the Sunbeam name and success which resulted in Mr K. Lee Guinness securing the much coveted trophy.

> At the same time they desire to express their sympathy with Mr Resta and Mr A. Lee Guinness for the mishaps which prevented them from finishing the race, but for which, they firmly believe

Sunbeams would have secured the proud positions of First, Second and Third.[28]

In an interview after the TT race with *The Motor*, Coatalen admitted that 'when one is endeavouring to win an important international event ... it must be recognised that the question of expense must not need too much consideration' which may have raised the eyebrows of his fellow directors a little but they were firmly behind his endeavours. He then went on to explain that before even designing the engine, the gears or the final drive ratio, it was necessary to thoroughly get to know the course for which it was intended. He also stressed how the drivers must be allowed to have their say and that they were largely responsible for the original design.[29]

Coatalen and his drivers had had very little time to get to know the course for the Grand Prix at Lyon as, when they had visited in May, they had been unable to drive at racing speeds, and by the time they got there in June there was only one official day of practice left so direct

---

28. Extract from Minutes of Directors' Meeting held on 16 June 1914 (Olive Coatalen album).
29. *The Motor*, 16 June 1914.

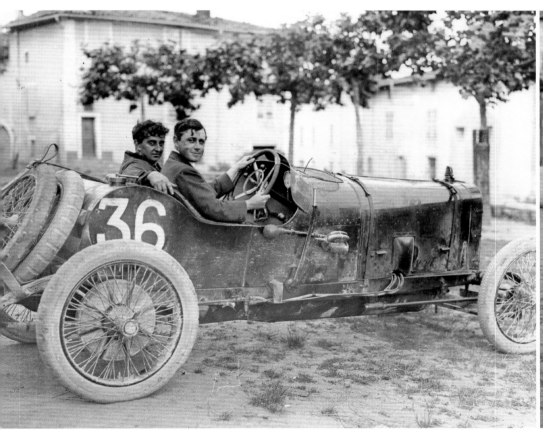

Coatalen with Cook in K.L. Guinness's 1914 Sunbeam in the village of St Maurice-sur-Dargoire during practice for the Grand Prix held at Lyon that year. Guinness retired on lap 10 with a broken piston.

Chapter Three

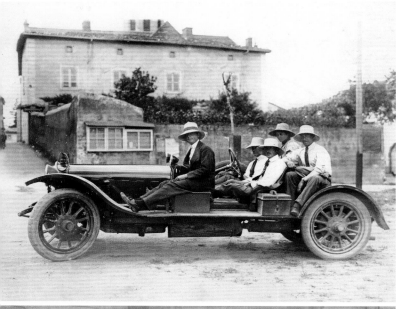

Top: Amongst the friends and supporters of the Sunbeam team present at the 1914 Grand Prix were Mr and Mrs van Raalte who had come in their own stripped-down six-cylinder Sunbeam. Suitably equipped with straw hats are Louis Coatalen (on tool box), Noel van Raalte (driver), Kenelm Lee Guinness (front passenger), Iris van Raalte and Dario Resta (back seats).

Bottom: Iris van Raalte appears to have brought some embroidery to keep herself occupied while the men played with the cars.

experience cannot have influenced the design of the Grand Prix cars very much. Certainly they had not been able to put in nearly as much preparation on the course as they had in the previous two years. Instead of a derelict château, the Sunbeam team headquarters for 1914 Grand Prix was at a little French inn with an obliterated sign over the door and a Union Jack fluttering above it in a small, inaccessible village, St Maurice-sur-Dargoire, some 20 miles south of Lyon not far from the back leg of the circuit. Chassagne, who had arrived slightly ahead of the others as soon as he got back from Indianapolis, used his racing car to transport baggage from Lyon railway station to St Maurice.

In contrast to the Tourist Trophy the competition was intense with forty-one entrants drawn from France, Italy, Germany, Great Britain, Belgium and Switzerland. As T.A.S.O. Mathieson wrote, it was 'a truly magnificent field … many consider it was the greatest race of all time'. At the start the Sunbeams accelerated away faster than the Delages against which they were paired but although they held the road well the British cars lacked the top speed of the fastest cars in the race, the Peugeots and Mercedes. These two teams indulged in a thrilling battle, which finally resulted in an overwhelming victory for Mercedes who took the first three places. Goux came in fourth on a Peugeot followed by Resta on the only surviving Sunbeam who must have felt satisfied at beating Rigal (Peugeot) who was seventh. (It will be remembered that Resta had led all the way in 1912 only to be beaten at the end by Rigal.)

On getting home from France, despite the gathering war clouds, several more competitive

Jean Chassagne and Louis Coatalen in Resta's Sunbeam for the 1914 Grand Prix. Chassagne retired on lap 13 with big-end failure. Resta finished fifth.

outings were squeezed in. Dario Resta drove *Toodles V*, the twelve-cylinder Sunbeam in the speed trials on Saltburn Sands, and Louis and Olive Coatalen were there with one of the Tourist Trophy race cars to witness his record attempts which were frustrated by strong winds. Kenelm Lee Guinness was also present and drove a Lyon Grand Prix car, winning his class. A week later his brother Algy was at the Leicester AC Hill Climb at Beacon Hill near Loughborough where he put up the fastest time with one of the Grand Prix cars. The final Brooklands meeting before the war was held on 3 August when Resta, again at the wheel of

Toodles V won his race. After that there was no more motor racing and everybody's priorities changed. Coatalen's experiments with aero-engines were soon to pay off, as is explained in the next chapter.

At the end of 1914, having recorded another record net profit for the Sunbeam Motor Car Company, Mr Thomas Cureton retired from his position of Managing Director, although he remained a director. In his place, Joint Managing Directors were appointed: Louis Coatalen, Chief Engineer and W.M. Iliff who had previously been Company Secretary, took on the responsibilities between them. C.B. Kay,

Chapter Three

who had been Coatalen's assistant since 1912, was appointed General Works Manager.

Other Business Interests

When Coatalen joined Sunbeam he brought with him experience of using Claudel-Hobson carburettors. Henri Claudel was a Frenchman known as 'the father of carburation' whose designs were imported and distributed in Britain through the business founded by Hamilton M. Hobson at Vauxhall Bridge Road, London. As the trade grew and it became worthwhile to make them in Britain, they were initially manufactured by a London instrument maker. Demand for these efficient carburettors soon became such that in 1911 it was decided that Hobsons should set up its own factory. By then Sunbeam was one of its best customers and with the help of Thomas Cureton, William Iliff and Louis Coatalen, premises were found in Cousins Street, Wolverhampton and a company, Accuracy Works Ltd, was set up in May. The three

Left: William Iliff, previously Company Secretary, was appointed Joint Managing Director with Coatalen in 1914.

Right: C.B. Kay, appointed General Works Manager of Sunbeam in 1914. He became a director in 1928.

Left: Accuracy Works Ltd, Wolverhampton, where Claudel Hobson carburettors were manufactured.
Centre: Major T.P. Searight.
Right: J. Montgomerie, who managed the Accuracy Works.

Olive Coatalen at the wheel, accompanied by Mme Claudel, wife of the carburettor designer.

Sunbeam directors joined Messrs. Hobson, Searight, Cheesman and de Poorter on the Board of what was in effect the manufacturing subsidiary of Claudel-Hobson. Turnover for the first year was £5,000, as Hobsons had given the Accuracy Works an agreement and a firm order for 5,000 carburettors.

Major T.P. Searight, who eventually became Chairman of H.M. Hobson Ltd, was often on hand during Sunbeam's record and racing activities in the early days of the company. On occasions Henri Claudel himself came to England and assisted with tuning Sunbeam engines. For Louis Coatalen these were not just business relationships but friendships based on shared passions.

Chapter Three

Private Life

If Coatalen failed for whatever reason to further his career by marrying a Miss Hillman, he soon made up for lost time by marrying a Miss 'Sunbeam' when he moved to Wolverhampton. Olive Bath was the only daughter of Henry James Bath (1867–1944), a Wolverhampton consulting engineer who was one of the original directors of the Sunbeam Motor Car Company when it was set up as a limited company in 1905 and who remained on the board until the merger with Darracq in 1920.

H.J. Bath (H.J.B.) was a great-grandson of the Cornish Quaker, Henry Bath (1776–1844) who, early in the nineteenth century moved to Swansea and, as a shipowner, established a major dynasty trading in copper with Chile. The second Henry Bath (1797–1864) – H.J.B's grandfather – further built up that business and also established the Swansea Iron Shipbuilding

A Bath family gathering in 1889 before the hunt. Henry James Bath is far left, his fiancée Olive Mary Griffiths stands on the other side of the doorway, second from right. They were married the following year.

Company in 1846. H.J.B.'s father, Edward Bath (1824–85, the second son of Henry Bath II) had worked in Chile and whilst there married Eugenie, one of the daughters of another important personage involved in that trade, Charles Lambert (1799–1876), a mining engineer whose family originated from Alsace (other members of the Bath and Lambert families also intermarried). When Edward returned to England in 1859 he managed his father-in-law's side of the business, as the Lamberts also had copper smelting works at Port Tennant, Swansea and Edward and his family lived in a large house nearby called Bryn-y-Mor. He died suddenly while addressing Swansea council in 1885.

H.J.B. was born in Swansea in 1867. He was Edward's fourth son (there were twelve children in all) and after being educated at Cheltenham appears to have worked in the family copper works for a number of years and one assumes this is where he learnt his engineering skills. In 1890 he married Olive Mary Griffiths, whose father Thomas Druslyn Griffiths (1837–1914) was another prominent figure in Swansea (President of the British Medical Association in 1903, a well-known surgeon but also 'a staunch Unionist and tariff reformer ... he took a

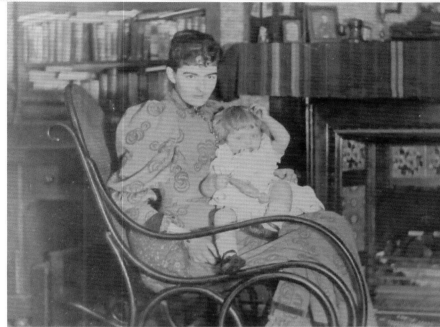

Olive Mary Bath (née Griffiths) with her daughter Olive Mary, aged two.

leading part in all public and philanthropic enterprises ... and was largely instrumental in the inauguration of the Swansea and South Wales Nursing Institute').[1] H.J.B. and Olive had a daughter, also called Olive Mary (who later fell in love with Louis Coatalen), who was born in 1891 and a son, Henry James, born the following year. They seem to have lived quite modestly in 5 Uplands Terrace, Swansea until H.J.B.'s mother died in 1896. For some reason after this they chose to move away from the polluted atmosphere of industrial Swansea to the beautiful, rural area to the west of Wolverhampton. In the 1901 census, when he was thirty-three years

---

1. *British Medical Journal*, Obituary Thomas Druslyn Griffiths, 25 April 1914, reprinted by the BMA in 1914. Incidentally, his wife Mrs Frances Annette Griffiths (née Gabb), was descended from a Franco-Swiss, Frederick Samuel Secretan, who had settled near Abergavenny, and as a girl had been despatched to spend a year in Switzerland and was thus fluent in French. After Dr Griffiths's death, she moved to the south of France with two of her other daughters.

Chapter Three

Pastel portrait of Olive aged ten-and-a-half by her aunt, the Swansea portrait painter, Gwenny Griffiths, in 1902.

old, he was described as a 'Retired Copper Works Manager' living at South Lodge, Norton near Bridgnorth.[2] Henry joined the army reserve and was a 2nd Lieutenant in the 3rd Battalion North Staffordshire Regiment Militia. As such, this battalion would not normally have been expected to serve overseas but the scale of the conflict in South Africa meant they were despatched to man block-houses in 1902 during the second Boer War.[3] H.J.B. was awarded a 'medal and two clasps' and left the army in 1904, the year before the Sunbeam Motor Car Company was founded.

Hervé Coatalen described his grandfather H.J.B. as a 'wonderful person' and 'very mechanically minded', although exactly what training he had in engineering is not known. He was undoubtedly a keen motorist, as in November 1903 he wrote a letter to *The Autocar* with a picture of his James & Browne 9 hp car that he had run on solid tyres since June 1902 and then in February 1905 he contributed an article to the same magazine about his experience of converting the car to pneumatic tyres. In this he noted that in the previous year he had used it on 262 days and covered 7,266 miles. His longest non-stop run during the year was from Wolverhampton to Swansea via the Brecon Beacons, a distance of 130 miles that had taken 6¾ hours.[4] At the Wolverhampton A.C. Hill Climb in September 1904, he took part on his 9 hp James Browne. Amongst the other competitors were no less than six Sunbeam entries and the drivers included J. Marston and T. Cureton, so H.J.B. would certainly have got to know the senior management of the Sunbeam firm at the time they were planning to incorporate it as a limited company. We know that he had a workshop in the attic of his house equipped with a couple of lathes where he spent a lot of

---

2. I am grateful to James Fack and Gill Loftus for digging out the census information.
3. Jeff Elson, Head of Research, Staffordshire Regiment Museum, 14 November 2014.
4. *The Autocar*, 11 February 1905, p. 206.

Louis and Olive Coatalen at Brooklands in 1910.

time making and mending things.[5] He was also an enthusiastic photographer who developed his own pictures. From these photographs it is evident that he led the life of a gentleman, hunting, shooting and fishing at various lesser country estates around the country. Having lived for a while at Badger's Heath, he had a new house built, not far away, at Ackleton, near Bridgnorth, in 1912. This house, The Folley, became the centre of family life and was where his grandchildren spent much time when they were growing up. H.J.B. died on 12 April 1944 and his wife died a year later on 29 May 1945.

Hervé Coatalen recounted the story that his mother Olive was reputed to have eloped with Louis by climbing down knotted sheets to escape from the family home. As they were married on 3 September 1909 at Birmingham Register Office and Olive gave her age as twenty-

5. Hervé Coatalen, *STD Journal*.

Winter sports party at Grindelwald during 1912. Olive sits on a sledge with Louis behind her in the white hat.

one years, when she was in fact still only seventeen (she was born on 11 October 1891) there may be some truth in the story. She was undoubtedly an enthusiastic Francophile before she met Louis, as she was already keeping diaries and notebooks in the language, so it is not surprising that she fell for his charm. Coatalen, in turn, seems to have been made welcome in Wolverhampton by the cosmopolitan Bath family and it is notable that, a few months after joining the firm, when he drove a Sunbeam in the Scottish Trials in June 1909, Henry J. Bath rode as his mechanic – an unusual position in which to find a director of the company. We know that Olive accompanied Louis when he went back to Brittany in August 1909 but whether other members of the family went with them on that occasion is not clear. Maybe she climbed down the sheets to go on holiday with him and after that he felt obliged to marry her. If it is true that Coatalen really thought he could advance his career by marrying the daughter of one of the directors, elopement seems a strange way to set about it. One might expect that this could have even hindered his career, but it does not appear to have done that either.

The 1911 census reveals that he and his

nineteen-year old wife were living in a nine-room house, The Orchard, 22 Pinefield Road, Wolverhampton. By the time their first son, Hervé, was born in September 1913 they had moved to Waverley House, Goldthorn Road, Wolverhampton, and then in 1916 with the family enlarged by a second son, Jean, they moved again, to Bromley House, Penn.

In the early years of their marriage Louis and Olive were an inseparable couple, with Olive accompanying her husband to race meetings at home and abroad, and even riding as his passenger when he took part in hill climbs. They were seen together everywhere from Brooklands to the Isle of Man and from Amiens to Monaco. They went winter sporting together at Grindelwald in Switzerland in 1912 and 1913 with groups of friends, and Louis and Dario Resta amongst others tried bob-sleighing as well as skiing.

Louis had been steadily climbing the social ladder for a number of years. The young 'orphan' from a Breton backwater had made his way through a mixture of skill and charm into a completely different milieu in England. Motoring was still very much a rich man's hobby and it was perhaps to his advantage that he was a foreigner with no recognisable class background. As soon as he started taking part in competitions he came into contact with affluent and influential people. Laurence Pomeroy's description of his own father's situation when he was made Works Manager of Vauxhall Motors in 1909, aged twenty-six, illustrates very well how things were at that time. Trained engineers were very thin on the ground and there were only 385 qualified members of the Institution of Automobile Engineers in 1908. 'A first-class man in the chief engineer's office could make a company's prosperity just as one with indifferent ideas or weak personality, could break it.' Pomeroy junior (writing in 1956) noted that, if the company was profitable, the engineer benefited from 'cash rewards on a scale hard to conceive in Europe at the present time'.

The social philosophy and negligible taxation of the Edwardian age made it possible for the successful man to both earn and enjoy a substantial differential above the average salary,

Christening party for Hervé Coatalen, born 15 September 1913, held at the Bath family home, The Folley, Ackleton. In the back row, left to right, are Henry Bath, Louis Coatalen, unknown (perhaps Searight), Kenelm Lee Guinness. Front row, left to right are Frances Griffiths (great-grandmother), Olive Coatalen (mother), Olive Bath (grandmother).

Chapter Three

and in L.H. Pomeroy's case had at the age of 24 a weekly wage of approximately twice that earned by the skilled mechanic, which was increased fivefold in five years. Thus at the age of thirty his net income was ten times that of the skilled weekly worker.... This had a technical as well as a personal and sociological importance. The chief engineers of the time lived on terms of financial equality with all but the richest of car users and could appreciate motoring from their customers' point of view.[6]

---

6. Laurence Pomeroy, *From Veteran to Vintage*, Temple Press, 1956, p. 90.

Olive Coatalen in another stylish hat at Brooklands. Algy Lee Guinness is on the left and Louis Coatalen on the right. Olive's outfit consisted of a black waterproof peaked cap with a cluster of cherries attached and a black and white check coat.

Signed photo of Olive Coatalen at the wheel of her 16 hp Sunbeam during World War I. It later suffered damp damage but has been photographically restored.

Louis Coatalen's situation was certainly similar to that of Pomeroy senior. His impact on Sunbeam's financial results was spectacular and he was personally able to share in that success. Pomeroy notes that although 'in 1909 Sunbeams made a profit of but £90, as the Coatalen programme gathered strength they rose to £20,700 in 1910 and £41,000 in 1911. After the great Sunbeam performance in the French Grand Prix of 1912 a profit of no less than £94,909 on a capital of £120,000 was realised in 1913.'[7]

It is perhaps an indication of his growing affluence that in 1914 he purchased from Gordon Pratt the 25-ft, wooden-hulled launch, *Tyreless IV*. Pratt had raced this boat, built by S.E. Saunders from a design by his own Cox & King studio, at Monaco in 1913 where it was not successful owing to an unsatisfactory Labor engine. Coatalen installed a six-cylinder Sunbeam engine instead. Although he cannot have used it much during the war it was still registered in his name in 1920.

---

7. Laurence Pomeroy, *From Veteran to Vintage*, p. 92.

Four

The First World War 1914–18

Breton becomes Briton

In retrospect Louis Coatalen's contribution to automobile history is probably more significant than his contribution to aero-engine history. It was, however, because of the aero-engines that he became a British citizen.

Aviation before World War I was very much in its infancy and, even if there was a growing band of enthusiastic pioneer aviators, the military possibilities of this new science were slow to be recognised in Britain, despite evidence that both Germany and France were investing in it heavily. It was not until 1912 that the British Government set out its plans for organising military flying, supposedly under a unified command but in practice split between the War Office and the Admiralty. The Naval Wing, which came to be known as the Royal Naval Air Service (RNAS), was fiercely independent – in part due to the fact that the First Lord of the Admiralty was a certain Winston Churchill, but also because it was run by remarkable men such as Captain Murray Sueter (later Rear Admiral Sir Murray Sueter) who was appointed Director of the Air Department. This independence was to work to the advantage of manufacturers like Sunbeam who subsequently became suppliers to the Admiralty, whilst the Royal Flying Corps, under Army control, sought to manufacture its own aircraft.

Events speeded up after the shooting of the Austrian Archduke Franz Ferdinand in Sarajevo on 28 June 1914. A month later the Austrians had declared war on Serbia and the British Navy (including its Air Arm) was put onto a war footing. Germany declared war on Russia on 1 August and on France on 3 August, the day they invaded Belgium; Britain declared war on Germany on 4 August.

On the very day that Germany declared war on France, Louis Coatalen went to the Wolverhampton solicitors, Fowler, Langley &

Wright, to complete the 'Humble Memorial' required to apply for naturalisation as a British citizen. He stated that he was 'a subject of the French Republic', that both his parents were French and had died, that he was a thirty-four year old Motor Engineer married with one son, 'Hervé Louis Coatalen aged 10 months'. The addresses at which he had lived during the previous five years were given as:

> April 1909 – December 1909: Grasmere, Oaks Crescent, Wolverhampton
>
> December 1909 – March 1910: Beacon House, Coalway Road, Wolverhampton
>
> March 1910 – September 1912: The Orchard, Merrivale Grove, Wolverhampton
>
> September 1912 – present: Waverley House, Goldthorn Road, Wolverhampton

The application was supported by six people whose signatures were obtained on 4 August. As well as getting support from pillars of the community such as a Commissioner for Oaths, F.J. Bell, and a Solicitor, Frank A. Stirk, the four other names were all intimately linked with the Sunbeam Motor Car Company. Edward Deansley, a Consulting Surgeon, of Claremont, Penn Road, was married to one of John Marston's daughters and was one of the original directors of the Sunbeam Motor Car Co., as was Samuel Bayliss of Claregate, Tettenhall, near Wolverhampton who gave his occupation as Iron Manufacturer and Justice of Peace. William Marklew Iliff, the long-standing Company Secretary and Director also vouched for Coatalen. However, perhaps most touching was the inclusion on the list of Thomas Herbert Harrison, whose job was given as Motor Car Manufacturers Manager, living at Merevale, Birches Barn Road, Pennfields. This was Tommy Harrison, chief racing mechanic, who had been with Coatalen from the very beginning of his competitive motoring exploits at Sunbeam and no doubt knew him better than most.

So far it is a fairly anodyne story of a Frenchman applying for British nationality, but subsequent events suggest that this was not an ordinary case and demonstrates the advantages of having friends in high places. The day after Fowler, Langley & Wright had submitted the formal application papers to the Home Office, they followed it up with a letter direct to HM Secretary of State in which they underlined the importance and urgency of Coatalen's case:

> We beg to make application that the consideration of the application may be expedited…. Mr Coatalen has lived in this country for fourteen years past … a Director of Sunbeam Motor Car Co., he attends to the manufacture of Aeroplane engines. We understand that the Admiralty and War Office require, or are likely to require, engines which will require Mr Coatalen's personal attention.

They drew attention to the fact that he was liable to be called upon to return to France to serve in the Army there although he had not yet

Chapter Four

received such a call. They concluded that he is 'married to an English lady and regards himself, in effect, as a British subject'.

The wires were soon buzzing as on the Home Office file there is a hand written note dated 6 August, which states:

> I understand someone at the Admiralty mentioned this case unofficially.... If a certificate is to be granted at all in this case it should be done as expeditiously as possible.... The grant of a certificate will not relieve the applicant from liability to French service but it may be of use to him for that purpose.

On 9 August the Home Office sent letters to Lord Kitchener at the War Office and to the Lords Commissioners of the Admiralty requesting any observations about Coatalen's application. The reply from the War Office effectively said 'no comment' but interestingly contains the following statement: 'I am commanded by the Army Council to inform you that inter-departmental arrangements have been made with the Admiralty whereby the latter are to have all the supplies of Sunbeam engines and the whole resources of the Sunbeam Motor Coy.'

Meanwhile Thomas Cureton, Sunbeam's Managing Director, wrote to the Member of Parliament for Wolverhampton, Mr A. Bird, on 8 August seeking his assistance with ensuring the 'essential papers be pushed through as quickly as possible in the Admiralty interest as we have received very important orders for Aeroplane Engines for them'. Alfred Bird (later Sir Alfred) had been Chairman and Managing Director of Alfred Bird & Sons Ltd (inventors of Bird's custard powder) before standing for Parliament and his link to Sunbeam was not simply because the works were in his constituency but he had been an enthusiastic pioneer motorist and vice-president of the RAC. Also his youngest son, Christopher, a motor-racing enthusiast, had driven Coatalen's aero-engined hill climb special at Shelsley Walsh in 1913, so was well acquainted with Louis Coatalen. Alfred Bird called personally at the Home Office on 10 August, showed them the letter from Cureton (which he left with them and is still preserved in the Home Office file) and asked that the case be put through as quickly as possible. He evidently spoke to the right people as a note was immediately despatched to the Admiralty requesting a response ASAP and it was returned the same day. At the bottom Captain Murray F. Sueter, Director of the Air Department, had written: 'If Mr Louis Coatalen became naturalized it will be a very good thing. He is most useful to us re-Sunbeam Engines.'

This seems to have been sufficient, as on 15 August Fowler, Langley & Wright was writing to the Home office thanking them for their letter 'informing us that the Secretary of State is prepared to grant a Certificate of Naturalization to Mr Louis H Coatalen'. The actual certificate (No. 25449) was issued on 20 August 1914, so the whole process had taken a mere three weeks, which is pretty remarkable.[1]

---

1. National Archives, naturalisation papers.

The First World War 1914–18

Left: An early side-valve V-8 aero-engine fitted with a propeller on a test rig.
Right: Side-valve V-8 engine mounted in a sub-frame and fitted with side radiators ready for installation in an airframe.

Aero-engines

As we have seen already, Coatalen's bright mind had been attracted to providing engines for aircraft well before a war in Europe was likely. Powered, heavier-than-air, flight was still a new activity although it was developing very rapidly. The first serious flight, a circular kilometre with controlled take-off and landing by Henri Farman on a Voisin, took place in January 1908 and Louis Blériot's cross-Channel flight followed in 1909. In 1911 the first non-stop flight from London to Paris (in just under four hours) took place and the following year Hubert Védrine flew for 100 miles in an hour and Roland Garros set a world altitude record of 13,200 feet. During the same period various countries began to use aircraft for military purposes.

Sunbeam's first 90 degree V8 (80 x 150 mm) intended as an aero-engine was under test before the end of 1912. Two of them were used in a racing boat in April 1913 and the Shelsley Walsh Special which Coatalen and Bird drove in June that year was powered by one of these engines by which time a 60 degree V12 (80 x 150 mm)

Chapter Four

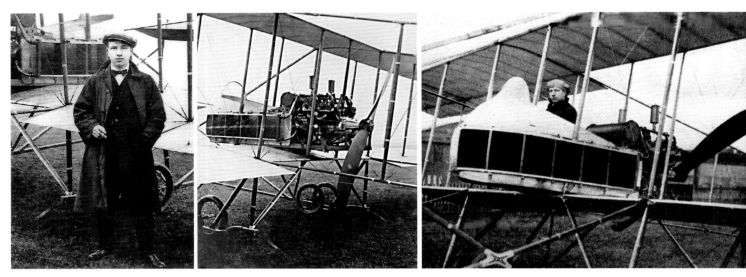

Left: Jack Alcock, later famous as the pilot of the first transatlantic crossing, was Sunbeam's test pilot before the war.
Centre: The Sunbeam V8 engine mounted behind the pilot to drive a 'pusher' propeller.
Right: Jack Alcock aboard the Sunbeam-powered Maurice Farman biplane.

was also already under development. According to Alec Brew (*Sunbeam-Aero Engines*), the first aeroplane to be fitted with one of the V8 engines was the Radley-England Waterplane entered for the 1913 Circuit of Britain Race but it did not participate even though Coatalen was present at Southampton to witness the start. That Louis Coatalen, as an individual, was the driving force behind the aero-engine developments is indicated by the report that, in order to test properly and further develop the engine, he acquired a Maurice Farman biplane.[2] Jack Alcock (later Sir Jack) was employed as the pilot to do the testing. The V8 engine was installed in the frame and was making its first flights by mid-October 1913. A replacement engine with larger bores was fitted in December, which Alcock demonstrated by taking up passengers from his base at Brooklands. During the first half of 1914 regular flights continued to be made, which included taking Coatalen himself as a passenger in February,[3] undertaking a half-hour night flight and a 50-mile circular flight at 5,000 feet in March, followed by participation in the London-Manchester-London Race in June. The War Office had purchased one of the 150 hp engines for testing in January. By July Alcock had clocked up a total of 30,000 miles of flying as a convincing demonstration of the viability of the V8 engine. In the meantime the V12 engine had been fitted in a car chassis, named 'Toodles V', and raced at Brooklands, although, because of the limitations of the tyres, it was only able to give the briefest indications of its potential.

By the time war broke out in August 1914 Sunbeam were relatively well positioned to respond to the country's needs and the

2. *NZ Motor & Cycle Journal*, 25 October 1918, p. 105. It seems more probable that the Sunbeam Company acquired the plane at his instigation.
3. Coatalen had already flown as a passenger in a two-seater Deperdussin plane in October 1911 at Brooklands (*Flight Magazine*).

Navy turned to the firm as one of their prime suppliers. Although initially engines were built in small numbers, by mid-August they were reported to have delivered 'in the region of three figures'[4] in response to a 'large order' from the Government. At the beginning of October *The Motor* was able to publish a photograph of the aviation engine department in full swing. As Alec Brew has written, 'The Sunbeam engines were the only high-power British engines available at the start of the war.'[5] Compared with the RAF 1 engine which was producing 10 horse power per litre in 1913 and had a power to weight ratio of 0.20 horse power per pound, Sunbeam's V8 developed 19 horse power per litre with a power to weight ratio of 0.24 hp/lb. But it soon became evident that this was not enough. Those responsible for aeroplane construction were constantly clamouring for more powerful engines. Sunbeam made minor improvements to their side-valve design, experimenting by moving the sparking plugs from their position between the valves to being above the pistons, but

Aero-engine assembly shop at the Sunbeam Works.

---

4. *The Motor,* 18 August 1914.
5. Alec Brew, *Sunbeam Aero-Engines,* Airlife, 1998, p. 16. Rolls-Royce started work designing an aero-engine at the beginning of the war. It first flew in December 1915.

returning them later to the original position, by which time watercooling had been introduced to the valve caps. Double valve springs were introduced as the long single springs proved inadequate. More obvious were the changes to the size of the cylinder bores which had increased from 80 mm to 90 mm in 1914 and then to 100 mm in 1915. The first two sizes simply made use of the boring equipment available for car production (the 12/16 car engine had an 80 mm bore and the 25/30 had a 90 mm bore) whilst 100 mm required a change for wartime production. This had the effect of raising the horsepower of the V8 from 150 hp to 160 hp and the V12 from 225 hp to 240 hp.

However, by 1915, the Navy was asking for a 300 hp engine to power aircraft capable of carrying bombs or torpedoes. Rather than starting with a blank sheet of paper, Coatalen and his team took their inspiration from their last racing car engine, the twin overhead camshaft 1914 GP Sunbeam. While the side-valve aero-engines had been based on automobile engines that had been well tested in both production cars and race cars over a number of years, the overhead camshaft design had only been run in a relatively small number of races. In the 1914 TT race, although K.L. Guinness won convincingly, the two other cars retired and similarly in the French Grand Prix that year, two of the three cars retired with engine problems (a big end and a broken piston). A little more experience had been gained by shipping two cars to America, where in 1915 racing was still taking place, but even one of these suffered piston seizure suggesting that lubrication was a problem.

Coatalen's approach to overcome the problem and to develop a powerful new engine was to adopt Ernest Henry's top end design with twin overhead camshafts for each cylinder block operating four valves per cylinder (modified to use pivoted finger cam followers). But the 'bottom end' was completely redesigned. The one-piece crankshaft was to have plain bearings between each cylinder rather than the ball bearings between pairs of cylinders for the two-part crankshaft of the racing engine. No less than three oil pumps provided adequate lubrication – one to scavenge the sump, one to supply the crankshaft and another small one to lubricate the camshafts.

Ever since W.O. Bentley had revealed the advantages of aluminium pistons when he joined the Royal Naval Volunteer Reserve (RNVR) in 1915, Sunbeam aero-engines had incorporated them, and the new overhead valve engines were no exception. To save some more weight the sides of the cast iron blocks were eliminated and aluminium panels were screwed on instead to form the water jackets. The side-valve engines had had the blocks, on either side of the vee, slightly staggered in order to allow two connecting rods to act side by side on the crankshaft with paired big-ends on each throw. In contrast, the new OHV engines incorporated another weight-saving idea which also enabled the length of the engine to be reduced slightly; they were fitted with articulated connecting rods so that two opposing cylinders acted on the same big-end bearing. This design feature, which Henry Royce also used on his Eagle engine having

The First World War 1914–18

Left: A V-12 twin overhead camshaft aero-engine fitted with carburettors but before fitting with exhaust manifolds.
Right, top: Partly dismantled V-12 Sunbeam Coatalen aero-engine showing the cascade of pinions to drive one pair of overhead camshafts.
Right, centre: Crankshaft, connecting rods, pistons, etc., of the V-12 aero engine.
Right, bottom: Camshaft casing, camshaft and articulated cam-followers.

adopted the idea from Renault V8 engines,[6] has been criticised for causing imbalances, but Coatalen, who was not shy of adopting better ideas if they were available and was never dogmatic about the superiority of his own ideas, persisted with this feature on all the V engines for the rest of the war, so he must have considered its advantages outweighed its disadvantages.[7] Even the booklet, *A Souvenir of Sunbeam Service 1899–1919*, published after the war, specifically mentions the articulated connecting rod arrangement as providing 'great gain as regards reduction of weight, amplitude of bearing surface, brevity and rigidity of crankshaft, and so forth'. From this design base, as well as an 18-litre, V12, 310 hp engine,

---
6. Derek Taulbut, *Eagle Henry Royce's first aero engine*, Rolls Royce Heritage Trust, 2011, p. 31.
7. *The Autocar*'s description of the early V8 motor boat engines (15 February 1913, p. 269) describes the connecting rods as being forked, so possibly Coatalen had tested that solution and concluded that articulated rods were better.

Chapter Four

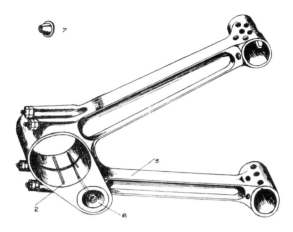

Articulated connecting rod.

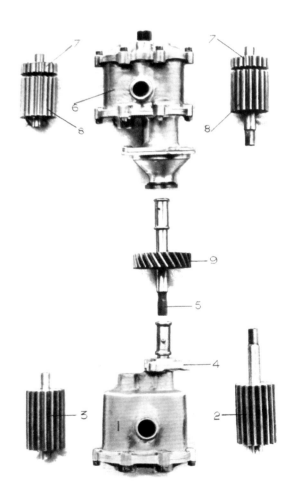

Oil pump assembly.

a smaller 200 hp, 11-litre, V12, a 9-litre straight-six and the extraordinary 33-litre, eighteen-cylinder (with three banks of six cylinders) producing 475 hp, were made available in 1916.

Although Coatalen had applied for a number of patents for himself and the Sunbeam Company in 1915, this activity reached a peak in 1916 as improvements to the twin overhead camshaft engines were made or planned. The lubrication of the articulated connecting rod was covered by one patent and another described an engine with an all-aluminium crankcase and cylinder block, flanged steel cylinder liners, detachable bronze valve seats and removable cover plates for access to the hinged finger-type valve followers, all features that would become familiar on the post-war racing engines. The aluminium engine was intended to have a detachable cylinder head but it is not thought that any were actually build with this feature. Many of these patents were also established in France, United States, Canada and Switzerland. Out of thirty patents taken out during the war, four of them were taken out jointly with H.C.M. Stevens, his Assistant Engineer.

Having demanded larger and larger engines, by early 1917 there had been a change of policy and the War Office was trying to standardise on engines of about 200 hp. At the heart of this was the Hispano Suiza 11.76-litre, 90 degree V8 engine HS8a, which the French had started testing in 1915 and which by 1917, after early problems had been overcome, with compression increased from 4.7 to 5.3, produced 180 hp (as HS8Ab). This was a much shorter engine (52 in long compared

The First World War 1914–18

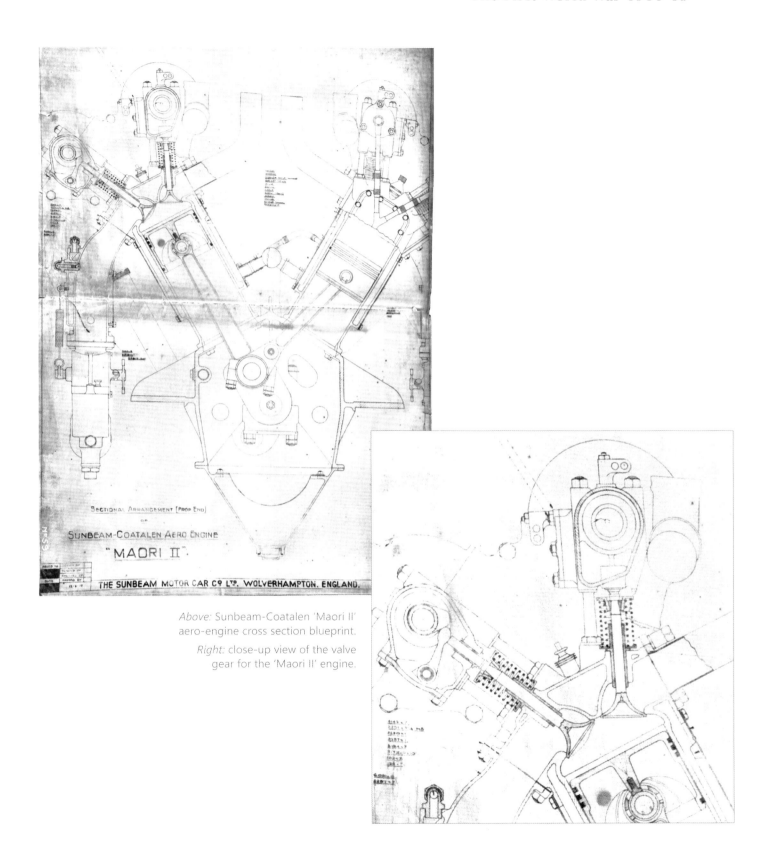

Above: Sunbeam-Coatalen 'Maori II' aero-engine cross section blueprint.

Right: close-up view of the valve gear for the 'Maori II' engine.

Chapter Four

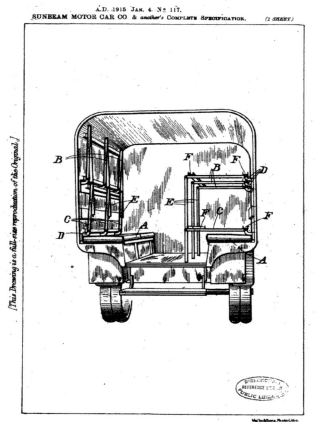
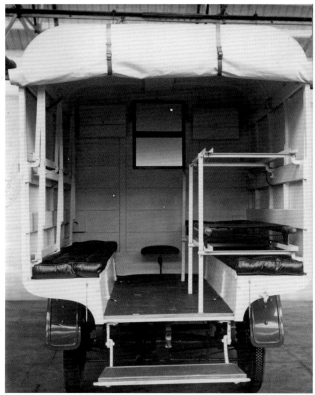

In 1915 one of Coatalen's earliest patents was for an improvement to ambulances that permitted stretcher racks to be folded away and the space used for carrying seated passengers.

to Sunbeam's V12 of similar capacity which was 75 in long) and much lighter (474 lbs compared to 950 lbs). The HS8 engine, despite suffering from vibration problems, would be built under licence in England in large numbers. Coatalen evidently thought he could 'improve' the design and built the very similar Sunbeam-Coatalen 'Arab' engine. He adopted cast aluminium for the cylinder blocks and the Hispano's layout of cylinder head with 3 valves per cylinder operated by a single camshaft per block which was shaft, rather than gear, driven. His modifications included pressed-in cylinder liners, rather than the original screwed-in liners, but he also replaced the forked connecting rods with articulated connecting rods which, combined with a redesigned crankcase that apparently gave less support to the main bearings, transmitted more vibration to the airframe.[8] On paper the 'Arab' looked an attractive proposition; it was only 41 in long, had a capacity of 12.26 litre and produced 208 hp for a weight of 550 lbs (including the propeller boss). This engine seems to have been developed very rapidly at the end of 1916 and in March 1917 an order for 1,000 (soon increased to 2,000) engines was received and arrangements were made for Austin, Lanchester,

---

8. Graham Mottram, *Sunbeam Aeroengines*, Fleet Air Arm Museum, 1984.

Napier and Willys Overland to start production of the 'Arab' under licence. But problems began to emerge with the engines under test, which were never satisfactorily resolved and fewer than 600 were actually built. Alec Brew summarised the saga:

> There is little doubt that the Sunbeam Arab was one of the worst engines of the First World War. The government made the huge mistake of ordering it in vast quantities after only a handful of hand-built prototypes had been bench run at the factory. They also then planned much of the country's aircraft production against the expected deliveries. Once the engine was tested in the air the terrible vibration problems which were to beset it came to light. These problems were never to be solved but attempts to do so held up production for many months. Perhaps this is the one good thing about the whole sorry saga. By the time the Arab

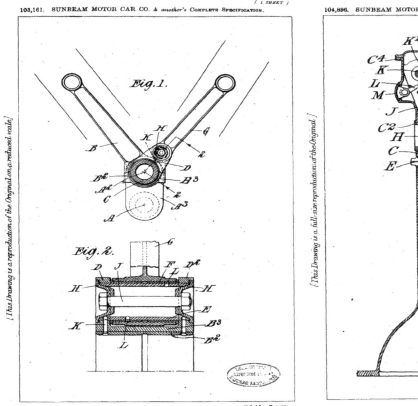
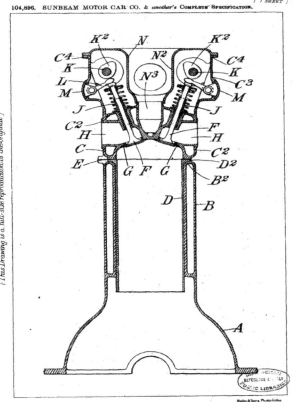

Left: Illustration from the patent concerning the lubrication of the articulated connecting rod design published in January 1917.
Right: Illustration from the patent for the design of an all-aluminium, twin overhead camshaft engine applied for in March 1916 and published a year later. It featured a detachable cylinder head with bronze seats for the inclined valves, central spark plugs, removable cam followers and dry cylinder liners. A later patent specified shrunk in liners and screwed in valve seats.

Chapter Four

The ill-fated 'Arab' engine.

was received in substantial numbers the war was almost over.⁹

In the meantime Coatalen was applying some of the features from this design (such as the aluminium cylinder blocks with steel liners) to engines developed from the existing twin overhead camshaft engines. In addition, towards the end of the war, he experimented with other ideas, such as replacing cast cylinder blocks with rows of individual cylinders bolted to crankcases and push-rod operated overhead valves, as used on Rolls-Royce and the Liberty engines. There were also 'fantastical' prototypes of a W 12 engine and even a massive twenty-cylinder engine that had five-cylinder blocks bolted in a star format around a common crankcase.

This plethora of designs, combined with the disaster of the 'Arab' engine, lead one to suspect that Coatalen was more taken up by the excitement of being involved with the leaders of the war at the highest level and enthusiastically

---

9. Alec Brew, *Sunbeam Aero-Engines*, p. 101.

responding to their wildest requests, than he was in seriously trying to improve the performance and reliability of the engines being produced. At the end of the war the Company justified the multiplicity of designs, saying, 'No aircraft engine manufacturing firm in the world has such a wide variety as the Sunbeam Company, for the simple reason that its pioneer work was so successful that, whenever a new type of power plant was had in desire by the authorities for some fresh form of aircraft, invariably the firm was asked to make its contribution to solving the problem.'[10] But the reputation of Sunbeam aero-engines for reliability at the end of World War I was not particularly good. How much responsibility for this should be laid at Coatalen's door is difficult to judge. His Assistant Engineer, who would have been in charge of producing the detail designs, was Herbert C.M. Stevens, who was credited with an 'ability to solve thorny engineering problems', and the Chief Draughtsman was L.J. Shorter, who had also joined the firm in 1914.[11]

In trying to push the boundaries of the output of these engines as rapidly as possible in a time of war, experiments had to be made with new designs and materials and then used in the unforgiving arena of flight. It might be considered that Coatalen's ability as a salesman was beginning to outweigh his ability as an engineer. However, most of the best engineers at the time did not yet understand the theoretical background to the vibrational forces that were being unleashed by these powerful new engines and so he was certainly not alone in being unable to eliminate all the problems encountered.

It is worth recording some of the more interesting applications of Sunbeam-Coatalen aero-engines. For example, they powered the Russian Sikorsky Ilya Muromets V biplane, which was the first four-engined bomber to be built. Subsequently, Sunbeam-Coatalen engines were used in the gigantic Handley Page Type O bomber intended for long-range bombing raids, which was the largest aircraft built in the UK at the time. In 1915 a Sunbeam-powered plane off Gallipoli launched the torpedo that sank the very first ship to be sunk by this method and in 1916 at the Battle of Jutland, which has been described as the greatest naval battle ever, a sea plane was used for reconnaissance for the very first time in naval warfare. Lieutenant Rutland, pilot of the Short 184 sea plane equipped with a Sunbeam Mohawk engine, launched from HMS *Engadine*, was able to radio back messages about enemy positions. Owing to low cloud it was necessary to fly at around 900 ft but four enemy light cruisers were identified, leading Admiral Sir John Jellicoe to conclude in his despatches that 'sea planes under such circumstances are of distinct value'.[12] Later Sunbeam engines were even used in boats as when Thorneycroft came

---

10. *A Souvenir of Sunbeam Service 1899–1919*, p. 28.
11. Herbert C.M. Stevens (1883–1948) moved around the industry a great deal during his career. His training had included spells with Napier and Thornycroft amongst others before he joined Sunbeam in 1914. He moved to the Darracq works in Suresnes, France, when STD Motors was formed after World War I, returning to England in 1923 as Chief Engineer for Talbot and Sunbeam. He joined General Motors in 1925, being appointed Chief Engineer of the Olds Motor Works, Lansing, Michigan, USA in July 1926. Stevens returned to England as designer and production engineer for Singer Co. in 1931 and then re-joined Sunbeam in June 1932 when he started work on the design of the Dawn. He later set up his own business, Avimo, supplying aircraft equipment to manufacturers.
12. Admiral Sir John Jellicoe, Despatch on Battle of Jutland, *London Gazette*, 4 July 1916.

Chapter Four

The gigantic Handley Page Type O long range bomber.

up with the idea of light, powerful, coastal motorboats they were powered by 450 hp 'Viking' Sunbeam-Coatalen engines.[13]

Coatalen used his contacts to seek to export his engines in a way other British manufacturers were less able to do. He even succeeded in getting his compatriot, Jean Chassagne, who had been called up by the French Army, reassigned to work on Sunbeam engines. In 1915 he had arranged for an engine to be tested in France and again more successfully a year later, but it was not until the end of the war that various French sea-plane makers flew planes powered by Sunbeam-Coatalen engines.

Policies and Public Duties

During the war Coatalen's position as the key figure at the Sunbeam factory was reinforced. When the Society of British Aircraft Constructors was created in March 1915 to set standards and pool information in the national interest, Louis Coatalen was appointed as Sunbeam's representative on that body. He was a member of its General Council and was elected to its fourteen-member Management Committee alongside people such as H.V. Roe of A.V. Roe and F.B. Parker of Short Brothers. Inevitably he was a member of the Society's Aircraft Engine Section but was also one of the four management committee members who put forward proposals for cooperation with the Aeronautical Society. The two societies reached an agreement to form a joint standing committee for mutual cooperation and support at the end of 1916. It was no doubt through these channels, as well as through his regular dealings direct with the Admiralty, that he came to know Lord Fisher, First Sea Lord (1841–1920) and Winston Churchill, First Lord of the Admiralty until 1915 (1874–1965), men for whom he would retain a life-long respect.

---

13. In 1917, Sunbeam-Coatalen aero-engines were retrospectively given 'tribal' names to reduce the confusion caused by simply referring to their horsepower, as had been done until then.

A French, Marcel Levy-Besson 'Deep Sea Patrol' flying boat triplane fitted with an eighteen-cylinder Sunbeam Coatalen 'Viking' engine.

Alphonse Tellier standing beside his flying boat biplane fitted with a Sunbeam Coatalen 'Cossack' engine, designed to carry three men, four 70 kg bombs and a machine gun.

In April 1915 it had been announced that the aeroplane engines being manufactured by the Sunbeam Motor Car Co. would in future be known as Sunbeam-Coatalen aircraft motors. Coatalen had managed to revive his dream of having his own name included as part of the 'marque'. This decision was implemented and the joint names appeared thereafter in advertisements and catalogues but there was a slight hiccup when it was sought to register the words 'Sunbeam-Coatalen' as a trademark in respect of 'internal combustion engines and parts thereof ... for aeroplanes and motor vehicles other than motor bicycles'. The Registrar of Trade Marks initially turned down the application principally on the grounds that there was confusion between the name 'Sunbeam' as applied to motor cars and to motor bicycles. The Sunbeam Motor Car Co. appealed and the matter was heard before Mr Justice Younger in the High Court of Justice, Chancery Division on 3 July 1916. In reviewing the evidence the judge referred to the fact that SMCC had been created to take over the business and goodwill of the motor car manufacturing activities of John Marston Ltd. Mr John Marston was still chairman and although he and his co-directors held a large block of shares they did not hold a controlling interest. Relations between Sunbeam Motor Car Co. and John Marston Ltd continued to be close but from the beginning the books of the two branches were always kept apart. There existed agreements between the two companies that SMCC would not manufacture motor bicycles and that Marston would not make motor cars. The public was not confused by the use of the same name on two very different products from two distinct businesses. The

Chapter Four

The Sunbeam-Coatalen Aero-Engine logo with a plane climbing through the sun's rays.

judge felt that the addition of the Coatalen name rather than creating more confusion would on the contrary make for a clearer distinction between the output of the two businesses. He noted: 'Mr Louis Coatalen is a prominent engineering official of the Sunbeam Company; he is an inventor, and the type of engine now turned out by the Company owes something to him.' It was concluded that the Registrar be directed to proceed with the application to register the trademark so the appeal was successful and the name Sunbeam-Coatalen was permitted to be applied to the aircraft engines made by the company.[14] The trademark provided for the name to be applied also to motor cars but as car production had ceased during wartime nothing came of this. The whole story is indicative of Louis Coatalen's standing within the company at the time. He was seen as the hero who had brought success and glory to the firm through racing which in turn had resulted in a greatly expanded business.

In May 1917 Louis Coatalen AFAeS (Associate Fellow of the Aeronautical Society) was invited to give a lecture to the Aeronautical Society at the Royal Society of Arts in John Street, London on the subject of 'Aircraft & Motor Car Engine Design'. In his speech he drew attention to the different requirements involved in developing engines for automobile production and for aero-engines for flight. In designing a car engine, he said, weight was not crucial but cost and the least amount of labour were vital, quietness was important, as was flexibility over a wide range of speeds; full power was seldom demanded but torque was needed from 300 rpm to 2,000 rpm. These attributes could be achieved with low compression ratios, small valve areas and simple sump lubrication. In contrast, minimising the weight of an aero-engine was paramount, whereas cost was not the only deciding factor provided the correct result was achieved: 'No machining is too expensive if it saves weight.' The aero-engine does not need to be quiet and does practically all its work at full power demanding maximum torque over a narrow speed range and a Brake Mean Effective Pressure (B.M.E.P.) of say 130 lbs (80 lbs for a car) per square inch. This implied the use of high compression ratios, large valve areas and strong valve springs, as well as dry sump lubrication to allow the oil to cool. The minimum horse power expected was 100 hp compared to 30 hp being usual for motor cars. Naturally enough Coatalen

14. Copy of transcript of the hearing held in STD Register archives.

underlined the similarities between designing aero-engines and engines for racing cars where reducing weight was important but the amount of time and labour used was secondary. Power had to be maximised and racing engines had to perform under full stress. Dry sump lubrication, now used in aero-engines, had been developed in motor racing as a response to the need to cool oil sufficiently when engines were being run at full throttle for long periods and offered the advantage of only keeping the oil under pressure for the relatively short distance between the pump and the bearings.

Coatalen concluded that engine design 'is all the time a question of compromise. The most successful designer is he who exercises the soundest judgement in weighing a hundred and one factors of the hour and who gives the shrewdest estimate of the relative value of each ... the business of the designer is to effect the best compromise possible.' He acknowledged that, while at the beginning of the war he had had the advantage of applying lessons learned from motor racing, the effect of war had been to accelerate advances and experimentation enormously so that 'the whole question of design is vastly more in a state of flux than the lay mind imagines'.[15]

Airships and Aeroplanes

In addition to building aero-engines, the Sunbeam factory was involved in the construction of complete aircraft during the war. These were mainly Short seaplanes but some Avro trainers were also built under licence. Coatalen's known policy of using the talents of the best people available is illustrated by the fact that F. Gordon Pratt was employed to set up aircraft production for Sunbeam. Pratt was a naval architect by training whom Coatalen had encountered when they had both raced motor boats at Monaco. This was not such a surprising appointment as it may sound, as there already existed plenty of links between boat builders and aircraft constructors on the Solent. For example, the shipbuilding firm John Samuel White had formed an aircraft department in 1913 and S.E. Saunders of Cowes undertook a joint venture with Sopwith Aviation to build a flying boat known as the *Bat Boat*, while Noel Pemberton Billing established his Supermarine factory to build 'boats that fly rather than aeroplanes that float' in September 1913.[16] Pratt joined Sunbeam in 1914, presumably soon after the outbreak of war, as manager of the aviation department and was responsible for building, organising and managing it. The first plane was delivered in November that year.

As Charles Faroux would remark many years later, Louis Coatalen harboured a secret love of big airships and many of the larger aero-engines were developed with airships in mind.[17] Possibly this interest in dirigibles stemmed from his having been in Paris during the time when Alberto Santos-Dumont had made the very first flights with an airship. This was in 1898, long before the first aeroplane flight, when Coatalen had just finished his

---

15. *Flight*, 17 May 1917.
16. Adrian B. Rance, *Sea Planes and Flying Boats of the Solent*, Southampton University Industrial Archeology Group, 1981.
17. Charles Faroux, 'Allocution', *Journal de la S.I.A.*, July 1953, p. 208.

Chapter Four

R. 34 airship undergoing a trial flight.

studies and got his first job in the French capital where the sight of an airship puttering across the sky, powered by a de Dion engine purchased from Coatalen's boss Adolphe Clément, must have been truly astonishing.

By 1916 Sunbeam-Coatalen 'Crusader' V8, side-valve aero-engines were being used to power twin-engine, non-rigid airships known as blimps on anti-submarine reconnaissance patrols over the North Sea, staying aloft for up to twelve hours. However, it was after the Armistice that some of the later engines found a new lease of life and fame powering the giant rigid airships. When the German Zeppelin L33 was brought down in Essex in September 1916, the government was made aware of how far in advance the German craft was in comparison to then current British designs; plans were made for building seventeen airships based on slight modifications to the Zeppelin. One of the differences was that each airship was to be powered by five 275 hp Sunbeam-Coatalen 'Maori' engines. Work started on constructing two airships, R33 and R34 towards the end of 1917 but by the time they were completed the war was over. They were enormous craft, measuring 645 ft (197 m) in length with a capacity of 1,950,000 cubic feet (55,218 cubic metres).[18]

Although they had been conceived to carry out bombing raids, protagonists for this means of transport lost no time in putting forward the potential for airships in peacetime. At a time when the carrying capacity and the duration of flights in aeroplanes were severely limited, airships, in contrast, were expected to be the liners of the future linking together the remotest places on earth. The Admiralty

18. In comparison a Boeing 747-8 Intercontinental (the largest Boeing) is 232 ft long (76 m).

agreed to lend R34 to the Air Ministry for long range flights and a trans-Atlantic crossing was planned as a fitting demonstration of its potential. In July 1919 it made the first ever east to west airborne crossing, as well as the first return journey by air, covering over 7,000 miles and taking seven days, and of course it was the first crossing by dirigible.[19]

Naturally enough the Sunbeam Motor Car Co. was extremely proud not only to have provided the power units to propel the R34 but also to have made the gearing, transmission and the propellers. The only mechanical problems encountered were a magneto that caught fire, broken valve springs, which were rapidly replaced by the engineers, and a leaking water jacket, which was staunched with chewing gum and copper sheet. However, on the return journey one of the engineers slipped and fell against the clutch lever of the starboard aft car while it was running under load, with the result that the engine raced and broke a connecting rod, so putting it out of action for the rest of the crossing.

In order to reduce fuel consumption, much of the outbound crossing had been undertaken with only two or three engines running. With all five engines turning at 1,600 rpm, an airspeed of 60 mph could be obtained and in good conditions, using only the two wing engines, she was still capable of 36 mph; petrol consumption was 'only' 25 gallons per hour. Not only was fuel consumption a problem (there were just 140 gallons left when they got to New York) but fundamentally the power of the 'Maori' engines left insufficient reserves for the craft if bad weather was encountered. On a test flight before the trans-Atlantic crossing it was found that R34 was unable to make any headway against a particularly strong adverse wind.

As a result, when the larger R36 (the first airship designed to carry passengers) and the even bigger R38 were launched in 1921, they were powered by 350 hp Sunbeam-Coatalen 'Cossack' engines and the number of engines used was increased from five to six, thereby augmenting the nominal power available from 1,375 to 2,100 hp.

The starboard forward car for R. 34 under construction at the Sunbeam works in Wolverhampton. The V-12 'Maori' engine is already fitted.

Financial Results

Generally the Sunbeam business seems to have fared well financially during the war but doing business was less straightforward. At the Annual General Meeting held in November 1915,

---

19. It was very nearly the first ever airborne crossing of the Atlantic but this prize was taken by Jack Alcock (by then a test pilot for Vickers) who, with Arthur Whitten Brown made a west to east flight with a modified Vickers Vimy bomber (powered by Rolls-Royce engines) in June 1919.

Chapter Four

John Marston was reported as saying that the company 'was now a controlled establishment' and since March 'their energies had been entirely at the service of the War Office, the Admiralty and the Allied Governments. They had supplied a very large quantity of cars and ambulances to the War Office and large numbers of aviation engines to the Admiralty. This aviation-engine had been the result of many month's experimenting by their brilliant engineer and designer, Mr Coatalen.'[20] In addition to supplying the Admiralty, 'quantities of cars' and aviation engines had been supplied to Russia.[21] France was also being supplied with aviation engines and a fully equipped

---

20. *Flight*, 19 November 1915, p. 908.
21. Getting paid by the Russian authorities was not simple and became virtually impossible after the Revolution; part payment of an invoice for some £18,000 issued in November 1918 was only received in 1989.

Above, top: Probably the main aft gondola for R. 34 being built at the Sunbeam works with the engines standing ready in the foreground. The aft gondola housed two Sunbeam Coatalen 'Maori' engines geared to one 20 ft propeller (STD Register archive).

Above, bottom: Four nacelles for R. 34 under construction with a Sunbeam Coatalen Maori engine installed.

Right: Letter of appreciation and congratulation from the Admiralty on the occasion of the safe return of R. 34.

```
C.Sec.S. 13093/19
                                        8th August 1919.

Gentlemen,
        I am commanded by My Lords Commissioners
of the Admiralty to express to you, on the occasion
of the safe return of H.M. Airship R.34 from her
voyage across the Atlantic, their cordial appreciation
of the skill and energy which your Firm devoted to
the construction of the machinery.
        My Lords would be glad if you would convey
Their sincere congratulations to all concerned.
                I am, Gentlemen,
                        Your obedient Servant,

Messrs Sunbeam Motor Car Co.,
  WOLVERHAMPTON.
```

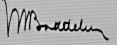

ambulance had been donated to the French Red Cross Society. Coatalen addressed the meeting for the first time as Joint Managing Director. He pointed out that not only had the policy of racing secured priceless publicity for the company but that the policy was further vindicated as the experience gained through developing racing engines had enabled them to supply aviation engines one month after war was declared. The Sunbeam Motor Car Co. had made 'a handsome profit' of £241,356.

A year later at the 1916 AGM the directors reported they were unable to present a valid balance sheet as the amount payable to the Government for excess profits under the Munitions for War Act had not been determined. However the volume of business done was greater than in any previous year and a 15 per cent dividend was agreed along with a payment of £5,000 to the staff bonus fund.[22] Finally, some months later, after agreeing the amount payable for two years' worth of Munitions Levy and Excess Profits Duty, the profit amounted to £50,025. The following year the figures were again delayed due to negotiations with the Ministry of Munitions and the Inland Revenue but the reported profit rose to £115,110. Thomas Cureton, Deputy Chairman who chaired the meeting due to the indisposition of John Marston JP, explained that working with a wide variety of government departments and the absence of coordination between those departments made life increasingly complicated. He cited the impact of wage increases awarded by the Government to the engineering trade that would represent an extra cost of £60,000 per annum for the firm. Mr Louis Coatalen, Chief Engineer and Joint Managing Director, was reported as stating to shareholders that the company's policy on munitions production was that of the military command – 'namely to be perpetually on the offensive and never to rest on their laurels'. For the year ending 31 August 1918 profits came to £92,264, so that the average annual profit during the war years worked out at £125,000. In January 1919, at an Extraordinary General Meeting it was agreed that it was necessary to increase the capital of the company by capitalising part of the reserve fund and by creating £80,000 of additional £1 shares.

22. *Flight*, 23 November 1916.

Chapter Four

Other Business Interests

One of Coatalen's good friends was Kenelm Lee Guinness, known as 'Bill'. This friendship stretched back at least to 1908 when Guinness had driven one of the Hillman-Coatalens in the Isle of Man Tourist Trophy Race. Guinness subsequently became a regular member of the Sunbeam racing team. He had run the 'Hilo' Ignition Ltd at Datchet with his brother Algernon and H.W. Bunbury, where they employed forty mechanics to repair and overhaul cars as well as making low tension magnetos and plugs, used on such vehicles as the 200 hp V8 Darracq. Guinness went on, after months of painstaking experiments, to perfect sparking plugs insulated with mica, which proved to be much better at resisting heat than the porcelain in use until then. He patented this design in 1913. 'Bill' set up production of K.L.G. Sparking Plugs at his yard behind what had been The Bald-Headed Stag pub in Kingston Vale and was manufacturing 4,000 to 5,000 per week by the beginning of the war in 1914. Demand for plugs for aero-engines was such that in 1916 the K.L.G. plug manufacturing business was turned into a limited liability company – Robinhood Engineering Works Ltd. Guinness's co-directors of the new company were Louis Coatalen, Noel van Raalte and Harold Percy, and they built a large factory on the site that increased production ten-fold, making two million plugs per year. Coatalen's involvement did not last very long, as he retired from the Board on 15 April 1918, but some years later he would take on the distribution and manufacturing of these plugs in France.[1]

Private Life

Coatalen's rise through British society continued during World War I. As already mentioned he had been invited to become a director of Sunbeam at the end of 1911 and by the early

---

1. Chris Ellis, *KLG From Cars to Concorde*, Smiths Industries plc, 1989.

Olive Coatalen in her nurse's uniform.

The First World War 1914–18

Top: Bromley House, Penn.
Bottom: Louis Coatalen with his two sons, Hervé (born 1913) and Jean (born 1916).

years of the war his role had become even more crucial so that when Thomas Cureton retired as Managing Director in December 1914, Coatalen became Joint Managing Director with W.M. Iliff. In this position he came into contact with many of those responsible for managing the war and could count the country's great leaders amongst his acquaintances.

Through the early years of the war, he and his wife Olive were living close to the Sunbeam Works at Waverley House, Goldthorn Road, Wolverhampton. Olive appears to have worked as a nurse, like her mother, as she was photographed in a nurse's uniform. In 1916, after the birth of their second son Jean, they moved into a larger house, Bromley House, in Penn, Wolverhampton. Despite the difficulties of war, they were still living life to the full, as far as possible. Coatalen was entertaining a party of six at Ciro's Club in London (a gourmet's dining club renowned for its African-American jazz band) when it was raided by police who suspected alcohol was being served after hours. Coatalen subsequently appeared as a witness for the defence to confirm, somewhat improbably, that 'no intoxicants were ordered or supplied after 9.00pm. The incident described by the police of a glass being swept off the table when they arrived was not intentional, the glass being knocked off through tilting of the table when the witness and his wife got up to dance. The chain bag which was stated to have been left behind by a lady belonged to his wife and was left on the chair by her while she was dancing. The glass of champagne found on his table had been poured out early in the evening for an officer, who would not drink it as he had just been inoculated.'[2]

The following year the family took a holiday on the van Raalte family's island, Brownsea, in Poole Harbour and there are

---

2. *The Times*, 15 December 1916. I am grateful to Luci Gosling for drawing my attention to this report.

Chapter Four

pictures of the children of both families playing together. Marcus Noel van Raalte's (1888–1940) main residence was at Bursledon, Hants, a little further east.

Although there are pictures of Louis with his new baby in the pram, he somehow does not look too comfortable in the role of doting father. In contrast, later that year Louis was pictured looking much more at ease, posing like a perfect English gentleman at a shooting party with his gun over his shoulder. Sitting next to him, similarly armed, was Iris van Raalte about whom there will be more in the next chapter. About a year later Iris and he were living together in a flat in Jermyn Street, London and Louis was separated from Olive. Olive continued to live at Bromley House for some months but the house and its contents were sold in March 1919.

The Coatalen and the van Raalte children playing together on Brownsea Island in 1917. Left to right: Gonda van Raalte, Charmian van Raalte, Jean Coatalen, Hervé Coatalen.

Shooting party, September 1917. Left to right: Olive Bath, Charteris, Louis Coatalen, Iris van Raalte, Noney, Olive Coatalen, T. Thornycroft.

Transatlantic Ties 1914–19

World War I appears to have awakened Coatalen's fascination with America and its possibilities, but on the whole his hopes of success across the Atlantic were frustrated time and time again.

Sunbeam had sent a racing car (*Toodles IV* rebuilt on a shorter chassis with a two-seater body) to the US to take part in the Indianapolis 500 Mile Race in 1913. In 1914 they sent a modified 1913 Grand Prix car, by which time four private owners had also acquired similar, six-cylinder Sunbeams to enter in various speedway races. Although the outbreak of war in August 1914 brought motor racing to a halt in Europe, the USA was not initially affected and so a couple of the 1914 Sunbeam Grand Prix cars were despatched to take part in the 1915 Indianapolis race to be driven by Noel van Raalte and Jean Porporato. The twelve-cylinder *Toodles V* was also raced in the US by Ralph de Palma and Hughie Hughes. These cars existed and were lying idle so it was perhaps not too difficult to justify these actions but it is more surprising that two more cars were sent to compete in 1916. The two cars that were prepared for the 1916 American racing season used 1914 Grand Prix chassis but were powered by brand new, 150 hp, twin overhead camshaft, six-cylinder 4.9-litre engines.

The fact that negotiations were in hand for Sunbeam-Coatalen aero-engines to be built in the USA may have had some influence on the decision. The Sterling Engine Co. of Buffalo, New York, despatched their Mr Homer to England on two or three occasions in late 1915 and early 1916 to negotiate an exclusive right to manufacture Sunbeam-Coatalen engines under licence, which were to be known as Sterling-Sunbeams. For this they paid Sunbeam $50,000 and received the drawings in late 1916. The Aircraft Production Division of the American Signal Corps ordered two 300 hp engines for testing and Sterling was given to understand that if these proved satisfactory large quantities of the engines would be required. However, when America joined the war in April 1917 and the decision was made to concentrate the nation's efforts on the Liberty engine, Sterling was told in August 1917 to forget about Sunbeam and prepare to manufacture Liberty parts.[23]

During the 1916 American race season the Sunbeam cars, equipped with the new 150 hp engines developed along the lines of the pre-war 'Peugeot' engines with twin overhead camshafts and four valves per cylinder but now in straight-six format and using aluminium pistons for the first time, were potent machines. In the hands of a number of different drivers at different circuits they achieved some respectable results, including four third places and a fourth place in the Indianapolis race (reduced to 300 miles in length that year) and this last result brought with it $3,000 in prize money.

It is not known if Coatalen personally visited the States at this period but it seems likely that he did because Eddie Rickenbacker, then racing for Maxwell, recounts in his autobiography that late in 1916 'I met a handsome, poised Englishman (sic) named

---

23. War Department Board of Contract Adjustment, Washington G.P.O., 1919, digitised by Google.

Louis Coatalen posing at the Sunbeam works, Wolverhampton, in one of the cars built to race in America in 1916. It was powered by a 4.9-litre, six-cylinder, 4-valve, twin overhead camshaft engine, fitted into the 1914 Grand Prix car chassis.

Louis Coatalen.' Coatalen invited Rickenbacker to join Sunbeam to develop cars to be raced in the USA during 1917, saying, 'I'm sure we could work out a mutually agreeable arrangement. We should like you to come to England at our expense and work with us in preparing the cars.' Rickenbacker sailed for England in December 1916 but upon landing at Liverpool was arrested on suspicion of being a German spy. For several days he was kept in detention aboard the boat *St Louis* on which he had made the crossing and was not allowed to set foot in England or communicate with Coatalen. Finally, he managed to telephone on Christmas Day and Coatalen was able to sort things out, arranging to meet at the Savoy Hotel in London the following day. By then Coatalen had obtained permission for Rickenbacker to return to Wolverhampton with him, provided he reported to the police station morning and night. The following day he was 'set to work once again designing and building a team of racing cars. As always, the work was hard, the hours were long but the job was interesting. The Sunbeam was an excellent model to begin with. The mechanics were efficient and helpful.' However, it was a short-lived project as early in February 1917

Germany declared unrestricted submarine warfare on the high seas and announced that Americans abroad had five days to return home. Rickenbacker set off full of ideas about creating an air squadron of fellow American racing drivers to join the war. However, he was back again in England in May as a sergeant in the US Army as part of a secret advance party on its way to France. His role was to be driver to Colonel Mitchell who had a Packard and later a Hudson (he sent Coatalen a photograph of himself with the latter car). Rickenbacker was only able to realise his ambition to become a fighter pilot in March 1918, but he had become America's most successful pilot by the end of war.[24] Before returning home he found time to visit Olive Coatalen in Wolverhampton where she photographed him in uniform at Bromley House in January 1919.

At the end of the war in November 1918, Europe was in no state to start organising major motor racing events but the management of the Indianapolis circuit was keen to promote the 500-Mile Race again and it was rapidly organised to take place on 31 May 1919 after only a two-year break. Based on the experience gained in 1916, the Belgian driver Josef Christiaens advised that a shorter chassis would improve handling on the bowl and was given the go-ahead to mount the six-cylinder, 4.9-litre engines in the 1914 Tourist Trophy race chassis. However, tragedy struck in February 1919 when he was demonstrating one of these newly rebuilt cars to the directors of the company outside the Sunbeam works; he hit a kerb, the car overturned, and he was killed.[25] Despite this setback two cars were despatched to America with Jean Chassagne and Dario Resta nominated as drivers. Chassagne, Resta and Tommy Harrison, Chief Mechanic, arrived at Indianapolis in early May just ahead of their cars and carried out a certain amount of testing with them after they had been prepared. About a week later, Louis Coatalen, who had also accompanied the cars and drivers across the Atlantic, was full of

24. Edward V. Rickenbacker, *Rickenbacker his autobiography*, Hutchinson, 1968, chapters 3-5.
25. Josef C. Christiaens was a Belgian racing driver and acrobatic aviator who had flown a Santos Dumont Demoiselle known as the 'Infuriated Grasshopper' before the war. He had been taken prisoner by the Germans but had escaped to England where he joined the Sunbeam Works.

Eddie Rickenbacker sent this signed picture of himself as a pilot to 'My dear friends, Mr & Mrs Coatalen'.

confidence and cabled that he was preparing the cars for an attempt on the one-hour record. However, on 20 May he suddenly announced that they were being withdrawn as their engines were slightly oversized. As this type of engine had run in 1916 and was later used in numerous competitions with their cubic capacity being declared as 4914 cc (within the 300 cu. in. limit) this has always been an unexplained mystery. It seems possible that the decision to withdraw might have had nothing to do with engine capacity but might in some way have been linked to problems with the Sterling Engine Co. At that time, it was claiming $168,482 in damages from the American Government in connection with its failure to obtain orders for Sunbeam engines, which it alleged it had been promised. Their complaint was not upheld by the US War Department Board of Contract Adjustment set up to consider such claims and, although they subsequently appealed to the Secretary of War, the Sterling Co. was not granted any compensation. It would no doubt have been feeling frustrated with Sunbeam and Coatalen in view of the expense it had incurred and the difficulty it had in constructing the first engines. Jack Hartley-Smith, a young Sunbeam mechanic, recalled 'we had the feeling at the works that "Coatie" had been threatened by someone for an infringement of something' so the cars returned home.[26]

During Coatalen's trip to the USA for the 1919 Indianapolis race it appears that he was tempted to invest in an American engineering business by a man called A. Eddie 'Chips' Smallwood with whom he had become friendly. By April 1920 he may have invested as much as $325,000 in shares of the US High Speed Steel & Tool Corporation. By the summer of that year he wanted to extricate himself but found he was locked in. Smallwood wrote on 9 September that he had had to oust 'Friedman' and 'Mac' because 'it looks like they have made a deliberate attempt to bankrupt the Company, as the affairs of the same are in a deplorable condition', but he tried to make it sound as though prospects were good. This was not what Louis wanted to hear, as he was convinced he already had a contract with 'the Pool' to buy back his shares in instalments. He insisted they had agreed to pay $10,000 immediately with the balance to be paid in twelve-monthly instalments. On 22 September he wrote back to Smallwood:

> I am very anxious to get some money, as you know I have resigned my position of Managing Director to the Sunbeam Co. and have no money coming in from that direction, and with half my capital in the steel shares producing no interest, I am very heavily hit and cannot wait.

Smallwood's reply, received nearly a month later, was obviously just playing for time as he tried to placate Louis, saying:

> I want you to know your interests in this Company are best preserved by playing along with us. We have acquired full control of the Company and we are going

---

26. Jack Hartley-Smith, letter to Anthony Heal, 21 January 1982 (author's collection).

> to make a big thing out of it. Just be patient and you will make more money out of this proposition than you ever dreamed of.

Coatalen was unmoved and continued to press for the deal to be honoured but he was still trying to recover his investment six months later. In the meantime, Eddie Smallwood had fallen ill as a result of the strain of trying to keep the business afloat and his position as President of the US High Speed Steel & Tool Co. had been taken by Marshal J. Root. Coatalen was then corresponding with Judge Emil E. Fuchs, who was supposedly representing his interests in the USA but was obviously closely linked with Smallwood. In early 1921 the situation did not sound promising as Fuchs wrote:

> I know of the deep affection and friendship that exists between Smallwood and yourself. I also know that Smallwood would rather lose everything he has than threaten your financial interest in this Company. He has surrounded himself with the best class of business ability ... and I believe with endurance and patience, on everybody's part, he will ultimately make a success of this Company.

It is not known how this matter was concluded although there was undoubtedly a meeting with both Smallwood and Fuchs when Louis Coatalen travelled to the United States in May 1921 for the Indianapolis 500 Mile Race. We do not know if Coatalen managed to get at least some of his money back.

The share deal was not the only dealing Coatalen had with Eddie Smallwood at this period as it also emerges that the latter had ordered a boat from the French boat-builder, Despujols (perhaps to be fitted with a Sunbeam-Coatalen aero-engine?). Coatalen wrote in one of his letters to Smallwood, at the end of December 1920: 'I am very hard up myself and have been obliged to suspend payment. Will you tell me what you intend doing ... as I must give definite instructions to Despujols (to cancel the order).' There seems also to have been some attempt at currency speculation with Louis buying Fr. 1m for Smallwood in August, which he sold again in December 1920 at a loss of $30,000 (about 23 per cent).

In addition, Smallwood was involved financially in backing Resta Motors in New York. Dario Resta, who had driven Sunbeam racing cars in England before the war, had since made a name for himself and earned a lot of prize money in the US at the wheel of a Peugeot, became the importer of Sunbeam motor cars to the United States in 1920. However, despite the fact that Sunbeams were the most expensive cars in America at the time, there were severe quality problems and Smallwood stated that one car had been returned after it had run 500 miles with the result that they had had to refund the client's money. The car Smallwood himself was running had required the work of two men for three weeks before it was useable. Coatalen agreed to despatch a competent mechanic to sort things out. Despite all these problems it appears that Coatalen and Smallwood did not fall out completely, as nearly ten years later

Chapter Four

Louis Coatalen standing in front of one of the 1921 8-cylinders Sunbeams at Indianpolis in 1921.

Smallwood's name occurs in Coatalen's notes on another investment opportunity he had taken up!

As will be recorded in the next chapter, Coatalen was back in America for the 1921 Indianapolis race for which three straight-eight engined cars had been entered under the 3-litre limit. Reflecting the new corporate structure of STD Motors Ltd, two were entered as Sunbeams while the third ran as a Talbot Darracq. That was the last time Coatalen flirted with racing in the USA and Resta's attempts at importing Sunbeams also petered out. For Coatalen, America had turned out to be the land of frustrated opportunities and it was to frustrate him again some years later.

Five

A New Beginning? 1919–23

The period after World War I might be considered to have started off full of promise and expectation for Louis Coatalen. He moved back to France a well-off, well-known, respected engineer with a new aristocratic woman by his side. In the years that followed, as a director of a newly formed international group of motor manufacturing companies, he was responsible for some highly successful racing cars.

However, it is equally possible to read the same events in a much less flattering light. His affair with Iris van Raalte may have jeopardised his acceptability in British society and perhaps it was partly this that lay behind his decision to move to Paris. Having left his wife and two children in order to live with another man's wife, his reputation in Wolverhampton would undoubtedly have hit rock bottom. The fact that he never received any acknowledgement from the British Government for his contributions to the war effort tends to confirm the hypothesis that he had suddenly become something of a social outcast.

Fortunately, his reputation as an engineer was still intact, based on his undeniable string of pre-war racing successes that were recognised on both sides of the Channel and he was able to build on these further by constructing cars to win important races and break the World Land Speed Record, but the anticipated commercial success did not follow.

Sunbeam Motor Car Company Ltd from 1918

The end of World War I coincided with major changes at the Sunbeam Company. John Marston, the founder and chairman, died in March 1918 at the age of eighty-one. He was a much-respected man in the town, being an Alderman and Justice of the Peace who had twice been elected Mayor of Wolverhampton. His wife only survived him by weeks and one of their sons had died shortly before. Marston was succeeded as Chairman by the man who had been at his right hand from the beginning, Thomas Cureton, but he had retired as Managing Director in 1914 and was himself unwell (he died in July 1921). The future direction of the business was therefore uncertain. A number of key members of staff moved as the end of war approached; L.J. Shorter, who had been Chief Draughtsman since 1914, and who had worked for Humber before that, took a similar job with Arrol-Johnston, and A. Mason, Assistant Works Manager, joined Guy Motors. But at a time when industrial relations were generally under strain with strikes and rumours of strikes being frequent, it is notable that at Sunbeam on the whole 'harmony prevailed'.[1] The directors gave a dinner for staff at the Queen's Café, Wolverhampton in March 1919 that was attended by 500 employees and friends, and entertained by the staff orchestra (Louis Coatalen was President of the Sunbeam (Moorfield) Orchestral Society). W.A. Priest, the Export Manager, proposed the health of the Joint Managing Directors W.M. Iliff and L.H. Coatalen, and 'dwelt upon the excellent relations between directors and staff'.[2] Iliff and Coatalen said they would do all they could to promote the interests of those employed. One notable new employee was Captain J.S. (Jack) Irving who joined the company in 1918 to take charge of the Experimental and Aviation department.

Coatalen was elected to the Council of the Institution of Automobile Engineers and things looked set to carry on as normal but only a couple of weeks later in April a bombshell dropped when it was announced that he had resigned his position as Joint Managing Director of the Sunbeam Motor Car Co. We do not know whether this was because of the uncertainty around the future of the firm or the result of his fall from grace caused by leaving his wife Olive, in order to live with Iris van Raalte. He remained a director and crucially it was stated that he would remain Chief Engineer, concentrating on design. The official line was that he had found it impracticable to give adequate time to both the increasing demands of the drawing office and the boardroom and so was to devote his attention wholly to design.[3] Very soon it emerged that it was his intention to do this from Paris.

It is notable that many of his contemporaries were honoured by the Government in recognition of their contribution to the war effort. Thus Herbert Austin received a knighthood while T.C. Pullinger (Arrol-Johnston), J. Harper Bean, A. Craig (Maudslay), J.G. Beardmore (Wm. Beardmore), E.M. May (Cubitt), A.A. Remington (Wolseley) all got CBEs and W.O. Bentley an MBE. Coatalen might reasonably have expected something. Although his daughter Marjolie believed he had been asked to pay for a knighthood and had refused, his son Hervé thought it more likely that it was simply that a double-divorcer could not be honoured at that time. There may be truth in both stories, as this was the period in which the Lloyd George Party Fund was

1. *The Auto*, 13 March 1919.
2. *The Autocar*, 29 March 1919, p. 468.
3. *The Auto*, 17 April 1919.

raising money by covertly 'selling' honours and Coatalen's personal situation might have allowed some pressure to be applied which he rejected. There is a suggestion in the contemporary press that his departure was due to him being 'peeved' at his treatment.[4] At the end of April, soon after the momentous announcement of his resignation had been made, he sailed to America on board the *Olympic* in order to accompany the two cars entered for the Indianapolis race and also to have a look at motor manufacturing and production over there. Coincidentally, his contemporary, Laurence Pomeroy, who was said to be pessimistic about the prospects of the motor industry, had left Vauxhall to work in the US just a month earlier, having been unable to agree terms for his own continued employment.

As previously mentioned, Dario Resta was appointed Sunbeam's representative in America and he opened a showroom in New York in conjunction with Eddie Smallwood before the end of the 1919. Coatalen was back in England again by mid-June and in September he was in Paris as rumours circulated that Kynochs was taking over Sunbeam. There was no truth in that rumour although Kynochs did buy a stake in John Marston Ltd, the sister company.[5] More significantly, in the light of future events, around the same time A. Darracq Co. (1905) Ltd bought out the entire share capital of Clément Talbot.

Top: Louis Coatalen taking William F. Bradley, Parisian correspondent of *The Autocar* for a ride in a new 24 hp, six-cylinder Sunbeam Sports Tourer in September 1919.

Bottom: Portrait of Louis Coatalen at the wheel of the six-cylinder Sunbeam in Paris, September 1919.

---

4. *The Auto*, 22 May 1919.
 It is now known that a knighthood could have been purchased for £10,000 and 1,500 were granted between 1916 and 1922. The OBE was said to have been introduced for those who could not afford a knighthood.
5. *The Motor*, 10 September 1919.

Chapter Five

During Coatalen's September visit to Paris he took *The Autocar*'s recently appointed continental correspondent, W.F. Bradley, for a run in the new 24 hp six-cylinder Sunbeam. This was when he revealed that it was his intention to establish an experimental department and drawing office for Sunbeam in the suburbs of Paris, for 'there is no place in the world like France for testing our cars and for carrying out experiments'.[6]

Meanwhile, Coatalen had purchased a large steam-powered yacht, *Zaneta*, from Montague Grahame-White for £7,000 after the latter had re-conditioned it following war service. It had originally been built by W. White & Sons, Cowes, in 1897, having a 102-ft long, steel-frame, planked hull. Previously owned by Lord Warsash, Grahame-White described it and its sister ship, as 'extremely pretty models,

both vessels having schooner bows' and being very well built.[7] Coatalen sailed the yacht to Paris and moored it at Longchamps, which was coincidentally immediately opposite the Darracq factory. In January 1920 the Seine flooded badly, causing many to be put out of work. W.F. Bradley reported: 'The person who is right in the centre of the floods and who can afford to be most indifferent to it is Mr Louis Coatalen…. The yacht is now completely isolated from land, for the lower portion of the Bois de Boulogne is flooded; but by means of a boat Mr Coatalen can always reach his residence.'[8] He sold the yacht back to Montague Grahame-White in 1923.

The Sunbeam Company had been inserting advertisements for Sunbeam-Coatalen aero-engines in the publication *La Vie Automobile* since August 1919, following the successful trans-Atlantic crossing by the airship R34, but the first physical sign of the firm moving its attention across the Channel was an imposing stand at the Paris Aero Show in early January 1920. This indicates that Coatalen and the other directors believed that the post-war prospects for aviation were good. A selection of Sunbeam-Coatalen aero-engines were displayed, including the monstrous 800 hp, 68-litre, V12 Sikh. The much less attractive commercial reality of the times was revealed when the Aircraft Disposal Department offered quantities (30,000 were reported to be available) of war surplus engines for sale. It was difficult, if not impossible, to sell new aero-engines at that time but efforts had to be made to continue to draw attention

Zaneta, Coatalen's first large yacht. The launch alongside appears to be one belonging to Kenelm Lee Guinness on a social call.

---

6. *The Autocar*, 20 September 1919, p. 444.
7. Lt Cmdr Montague Grahame-White, RNVR, *At the Wheel Ashore and Afloat*, Foulis, 1934.
8. *The Autocar*, 17 January 1920, p. 117.

A New Beginning? 1919–23

Left: Sunbeam Despujols I at Monaco in 1920. Powered by a 425 hp, Sunbeam-Coatalen 'Matabele' V12 aero-engine, it was timed at 75 mph on the Seine and took first place in the 600 km Lyon to Monaco event driven by the Marquis de Soriano.
Right: Louis Coatalen and Victor Despujols at the latter's boatyard.

to their capabilities. In the spring of 1920 the 350 hp Sunbeam racing car was developed in Wolverhampton with an eye to taking records at home and abroad in the absence of a formal racing programme. At the same time Sunbeam-Coatalen aero-engines (425 hp 'Matabele') were installed in several racing boat hulls by the Despujols boatyard on the river Seine. One of these boats achieved 120 km/h in trials on the river and won several races at Monaco in April.

During January 1920 Coatalen set up Sunbeam's experimental drawing office at 53 rue de Colombes, Courbevoie, the Parisian suburb next to Suresnes where much of the automobile industry was concentrated. It seems probable that Herbert C.M. Stevens, who had worked in the drawing office at Wolverhampton on aero-engine design and described himself as 'chief designer and assistant engineer', was among the first to work there having moved to Paris alongside Coatalen.

The Creation of STD Motors, 1920

In June 1920 news of the proposed merger between the Sunbeam Motor Car Company and A. Darracq Co. became common knowledge. What Coatalen's role in the negotiations had been is not known but the outcome of the merger was that he was firmly back on board as a director of the new holding company, STD Motors, and also appears to have regained his function as Joint Managing Director of Sunbeam. He was to be based principally in Paris and his main role seems to have been to assume responsibility for the competition affairs of the whole STD Group, a task he set about with renewed enthusiasm. Designs for a new racing car with a 3-litre, straight-eight, twin overhead camshaft, 32-valve engine in a new chassis were soon put in hand in readiness for the French Grand Prix to be held in July 1921.

STD Motors

STD Motors consisted of three motor manufacturing concerns, Sunbeam, Talbot and Darracq, plus a number of other linked businesses. The growth of the Sunbeam company has been outlined already. The other two car marques also had stemmed from liaisons with French car makers. Clément-Talbot had been founded in 1902 in North Kensington by a French entrepreneur, Adolphe Clément (Louis Coatalen's old boss) and the Earl of Shrewsbury and Talbot, to produce in England automobiles to the design of Monsieur Clément. Over the years purely English Talbot designs evolved but at the end of the war the Earl wished to recover his investment in the firm and sold out to A. Darracq & Co.

The Darracq company, started before the turn of the century, was the outcome of the energies of another French entrepreneur, Alexandre Darracq. In 1902, in order to raise a large amount of capital with the aim of producing a lot of cars cheaply, Darracq floated a company in London that would own his factory in Suresnes outside Paris where the cars were made. In 1905 a new company was formed to provide even more capital (A. Darracq & Co (1905) Ltd), which absorbed the old company and continued to grow until things went wrong in 1911. Alexandre Darracq retired from the position of Managing Director and the English directors then appointed Owen Clegg to manage the business in France. At the end of World War I there was an English subsidiary, Darracq Motor Engineering, as well as the French factory and predominant interests were acquired in the engineering businesses, Heenan & Froude, and Jonas Woodhead, the vehicle spring makers. In 1920 the holding company A. Darracq & Co. (1905) Ltd amalgamated with the Sunbeam Motor Co. Ltd. and at the same time acquired W & G Du Cros Ltd of Acton; the new group was renamed STD Motors Ltd. The cars made in France were then called Talbot Darracqs until 1922 when that subsidiary was renamed Automobiles Talbot SA and they became Talbots except when they were sold into the UK where they had to revert to being Darracqs to avoid confusion with the English Talbots. For the sake of clarity throughout this book the cars made in Suresnes are referred to as Talbot-Darracqs.

In October René Thomas appeared at the Gaillon Hill Climb, south-east of Rouen, at the wheel of the 350 hp, V-twelve Sunbeam, to break his own record of the previous year, at 108.6 mph. It was powered by an engine that was built especially for the car but incorporated much wartime aero-engine experience, such as aluminium cylinder blocks with steel liners, aluminium pistons and articulated connecting rods. The 350 hp Sunbeam was destined to have a long career and become an iconic record breaker.

1921: Motor Racing Again

Whilst the production of touring cars for sale to the public was to continue largely independently at the three factories in Wolverhampton (Sunbeam), London (Talbot) and Suresnes (Talbot-Darracq), Coatalen initially sought to simplify racing car design demands by having one model that could be fitted with different radiators to represent all three of the manufacturers forming the STD group, an early example of a technique later known as badge-engineering. It implied building racing cars on a scale he had not previously undertaken as no less than ten of the cars were needed – three for Indianapolis in May and seven for the French Grand Prix in July. For Indianapolis, two would run as Sunbeams and one as a Talbot-Darracq. For the Grand Prix, three were entered as Talbot-Darracqs (French), two as Sunbeams (English) and two as Talbots (English). It seems probable that the engines were all built in Wolverhampton and then distributed to the other factories to be assembled on the chassis. By the end of March Segrave was testing one of the cars at Brooklands and won a race with it at the Easter meeting before three were shipped out to America in mid-April. The Anglo-French nature of this development was underlined by the fact that the mechanic assisting Segrave was a young Albert Divo who had come over from France. When the team sailed for the USA, Divo was one of the mechanics alongside two French drivers, André Boillot (who was to drive the Talbot-Darracq) and René Thomas (who would handle one of the Sunbeams) alongside the other Sunbeam driver, Dario Resta, with Coatalen in charge of the team. At the Indianapolis 500 Mile Race the cars showed

René Thomas at the wheel of the 350 hp Sunbeam at Gaillon Hill Climb, south of Rouen, in October 1920. He set a new record for the course at 108.6 mph.

Chapter Five

Top: One of the 3-litre, eight-cylinder cars at Indianapolis, 1921. Note the upward slope of the exhaust pipe which reflects the angle of the engine in the frame and the additional oil-cooling pipes below. Standing behind the car are Dario Resta, Ora Haibe and Louis Coatalen. In the driver's seat is perhaps Eddie Rickenbacker.

Bottom: The STD team at Indianapolis, May 1921. Standing, left to right: André Boillot, Capt. Irving(?), René Thomas, Frank Bill, Dario Resta, Louis Coatalen, Eddie Rickenbacker(?), Albert Divo.
Seated, left to right: unknown (Boillot's mechanic), unknown, unknown, Ora Haibe, unknown, unknown.

that they were competitive if not quite as fast as the Ballot team from France; René Thomas held third place until the 144th lap when a water connection broke, André Boillot was unfortunate when a piece of broken gudgeon pin circlip jammed his oil pump on the 41st lap, but Ora Haibe, a young local driver who had stepped in to take the place of Resta who was ill, drove a steady race in the third car to finish fifth.

While the team was away in America, work was supposed to continue for the preparation of the other cars for the French Grand Prix. In particular they were to be fitted with front wheel brakes which was a first for a Sunbeam racing car and which had not been considered necessary for Indianapolis. Irving wrote from Wolverhampton to Coatalen, who had just arrived at Indianapolis, explaining that he and the designer Mr Stevens had been into the matter thoroughly with Kenelm Lee Guinness and Segrave, and that the only satisfactory arrangement was to use the hand lever to apply the front brakes as the pedal travel could not provide sufficient leverage.[9]

By the time Coatalen arrived back in Europe there were only six weeks until the Grand Prix and it was obvious that the management structure of the new STD Syndicate was not yet properly functional. Possibly too many details could only be decided by Coatalen himself,

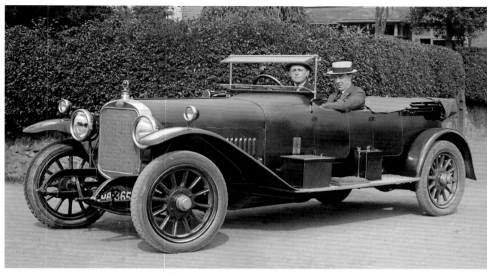

Louis Coatalen, the Sunbeam designer, and Owen Clegg, the Darracq designer, both became directors of the STD Combine when it was created in 1920. They are seen here in a 24 hp Sunbeam.

but the cars were very far from ready for other reasons as W.F. Bradley recalled in his memoirs:

> Daily visits to the Talbot-Darracq factory revealed a state of chaos. We had witnessed a real fight between Clément, in charge of the engine test bench, and Gaumont, responsible for chassis assembly. The workmen stood behind their respective chiefs and were much more interested in seeing a renewal of the fight than in preparing the racers.... Owen Clegg, as works manager, expressed his disgust for everything connected with racing in a truly Yorkshire manner.[10]

It seems that it was on Clegg's advice that the STD directorate on 16 July took the decision

---

9. Letter from J.S. Irving to L. Coatalen, 12 May 1921 (author's collection).
10. W.F. Bradley, *Motor Racing Memories*, Motor Racing Publications, 1960, p. 171.

1921 Grand Prix de l'A.C.F. at Le Mans. K.L. Guinness in the pits for one of nine tyre changes during the race. His 3-litre, eight-cylinder Talbot finished in eighth place.

to withdraw the entire team from the race.[11] With little more than a week left before the race, Segrave and Guinness, who were appalled and indignant about the decision to withdraw, 'remonstrated with Coatalen and Clegg but both said they were powerless to alter the decision of the directorate'.[12] So, the two drivers rushed back from France, where they had been practising, to persuade the board to change their minds, even offering to bear the running costs. They tackled Arthur Huntley Walker, Managing Director of the group, and finally and reluctantly he was persuaded that a partial team was better than no team at all and would do less damage to the firm's reputation than an ignominious withdrawal. With this agreement, according to Bradley, 'two young Englishmen, who had no connection with the racing department (one was from the buying department, the other from accounts) ... placed themselves at the head of the organisation, ordered the foremen to cease quarrelling and

---

11. See letter to *The Autocar* from Malcolm Campbell, 23 July 1921.
12. Cyril Posthumus, *Sir Henry Segrave*, Batsford, 1961, p. 63.

infused such enthusiasm among the workmen that the cars were finished in time to go to the starting line at Le Mans'. So it was that two Talbots (Segrave and Guinness) and two Talbot-Darraqs (René Thomas and Andre Boillot) started the race. The fact that no Sunbeams were there was explained away as being due to the difficulties caused by the miners' strike.[13]

In contrast to the smooth surface of the 'Brickyard' at Indianapolis, the circuit at Le Mans consisted largely of loose stones which caused much damage to the racing cars and the competitors themselves. Although René Thomas had a succession of mechanical difficulties, it was a stone through the oil tank which put him out of the race. Guinness had to whittle a wooden plug to replace the petrol filter tap knocked off by a stone and Segrave's mechanic, Moriceau, was briefly knocked out by a similar flying object. But the other main problem encountered by the team was that their Dunlop tyres were not up to the task. In contrast to the Ballots equipped with Pirelli tyres which ran throughout the race without needing to be changed, Boillot had to change nine and Segrave fourteen tyres. At the end of the race Boillot was placed fifth, Guinness seventh and Segrave struggled to finish in eighth place on just six cylinders.

In reviewing the lessons to be learned from the Grand Prix, the report in *The Autocar* said of the STD team 'that there was insufficient organisation, the old phrase *ordre, contreordre, désordre,* applying with extraordinary aptness'.[14] In retrospect it seems probable that the underlying reason for the debacle in the preparation of the 1921 Grand Prix cars stemmed in part from a clash of personalities between Louis Coatalen and Owen Clegg. Clegg had managed the British-owned, Darracq factory in Suresnes since 1912. He was answerable to his co-directors in London but for day-to-day operations was solely responsible and his character was such that he would not tolerate interference. An article about him published in *The Motor* in 1914, which mentions his strong personality, great ability and shrewd common sense, gives an indication of his no-nonsense attitude to management:

> To have an Englishman placed in authority over them would be no pleasure to the French staff, but I fancy their chagrin would in no way disturb Mr Clegg. A Yorkshireman, with the shrewd common-sense appreciation of the things that matter peculiar to the natives of the big county, he would very likely treat sentimental considerations with indifference, but I can scarcely fancy anyone enjoying the experience who had the hardihood to indulge in open opposition.[15]

For such a man to find that a 'cuckoo' with the very different personality of Louis Coatalen had

---

13. According to G.R.N. Minchin, *Under my Bonnet*, Foulis, 1950, Coatalen also returned to England around the same time as Guinness and Segrave. After the last-minute negotiations had taken place Minchin and Phil Paddon travelled back to France with Segrave and Coatalen, crossing from Southampton to St Malo. They spent the night there and lost all the money they had with them at the casino before travelling on to Le Mans.
14. *The Autocar*, 6 August 1921.
15. *The Motor*, 19 May 1914, p. 720.

Chapter Five

Left: The twin overhead camshaft, 16 valve, 1500 cc engine of the 1921 Talbot-Darracq.

Below: Victorious on their first outing. The Talbot-Darracq team after the 1921 Grand Prix des Voiturettes held at Le Mans. Left to right: H.O.D. Segrave and J. Moriceau (third place), R. Thomas and A. Divo (first place) K.L. Guinness and unknown (second place).

landed in his nest must have been very difficult to accept. Their methods were different and Clegg disapproved of the huge expenditure on racing cars. He must have been concerned that this Frenchman with natural charm would undermine his position as the authoritarian outsider. Clegg's command of French was limited, and he was dependant on a bi-lingual secretary to communicate his instructions. The two men had to establish some lines in the sand if they were to work together. It is noteworthy that subsequently entries for the Talbot-Darracq racing cars at Brooklands were made by Mr Huntley Walker, not Louis Coatalen, and journalists describing the cars regularly mentioned they were designed by Coatalen but built 'under the supervision of Mr Owen Clegg'.

Whatever internal changes were made they were extremely effective. The new 1½-litre Talbot-Darracq racers (essentially scaled down versions of the Grand Prix cars – using half the engine), unveiled at the beginning of September and said to have been produced in twelve weeks, proved to be extraordinarily successful right from the very outset. According to *The Autocar*, the engines were British-made and fitted into the chassis in France. In the Grand Prix des Voiturettes race held at Le Mans only a few weeks later they swept to absolute victory with Thomas finishing first, closely followed by his team mates Guinness and Segrave. For the 279 mile race the winner averaged 72 mph whereas the winner of the Grand Prix with an engine of twice the capacity had averaged 78 mph. One notable change compared to the STD 3-litre cars was that the 1½-litre cars used Pirelli tyres. Lessons had been learned.

Top: The Talbot-Darracq team for the 1921 Brooklands 200 Mile Race pictured outside the K.L.G. Sparking Plug Works, fitted with streamlined nose cowls. No 69 K. Lee Guinness and A. Divo, no. 50 driven in race by J. Chassagne and Cozens, no. 33 H.O.D. Segrave and J. Moriceau.

Bottom: Segrave and Moriceau with the 1½-litre, four-cylinder, 16 valve Talbot-Darracq in which they won the 1921 200 Mile Race at 88.8 mph. Note the streamlined body by Hawkers.

Chapter Five

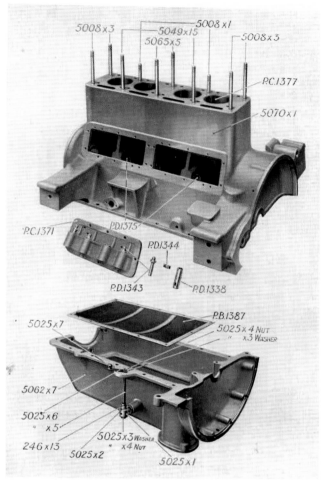

Cast in-one aluminium cylinder block and crankcase for the 14 hp Sunbeam introduced in autumn 1921.

From Le Mans the cars went straight to England (except Thomas's car which made a brief detour to take part in the Gaillon Hill Climb) where the little Talbot-Darracqs reappeared in October for the 200 Mile Race at Brooklands for which they were fitted with more streamlined bodies built by Hawkers. Again, the cars ran like clockwork, finishing in the first three places, and this time it was Segrave who won, followed very closely by Guinness. In third place was Malcolm Campbell driving instead of Thomas. Segrave averaged just under 89 mph from start to finish and his fastest lap was 97.65 mph. The Bugattis, which had been expected to provide opposition, were outpaced.

After the race Louis Coatalen invited the drivers and mechanics, along with several prominent members of the automobile world, to 'a very cheery little dinner' at the Criterion Restaurant in London to celebrate this remarkable victory and to close the racing season on a high note.

At the end of the war it became apparent that Sunbeam had not prepared designs for any new production models. Initially they had to make do with updated versions of the side-valve 12-16, renamed 16 hp, and the 24 hp which was a partially redesigned version of the 25-30 with a reduced capacity engine. Behind the scenes it appears that the new centralised design office was slowly getting into its stride but it was not until October 1921, more than a year after the merger and two years after the end of war, that important changes were announced to the production car ranges for the 1922 season. Sunbeam and Talbot both introduced identical new 14 hp models with an integral aluminium cylinder block and crankcase as well as novel pear-shaped combustion chambers (Owen Clegg had visited Wolverhampton in May and been impressed at the record speed with which Works Manager, C.B. Kay had got the prototype on the road),[16] and the engines of the 16 hp and 24 hp cars were converted to overhead valve operation.

---

16. Letter to L. Coatalen from J. Irving, 12 May 1921.

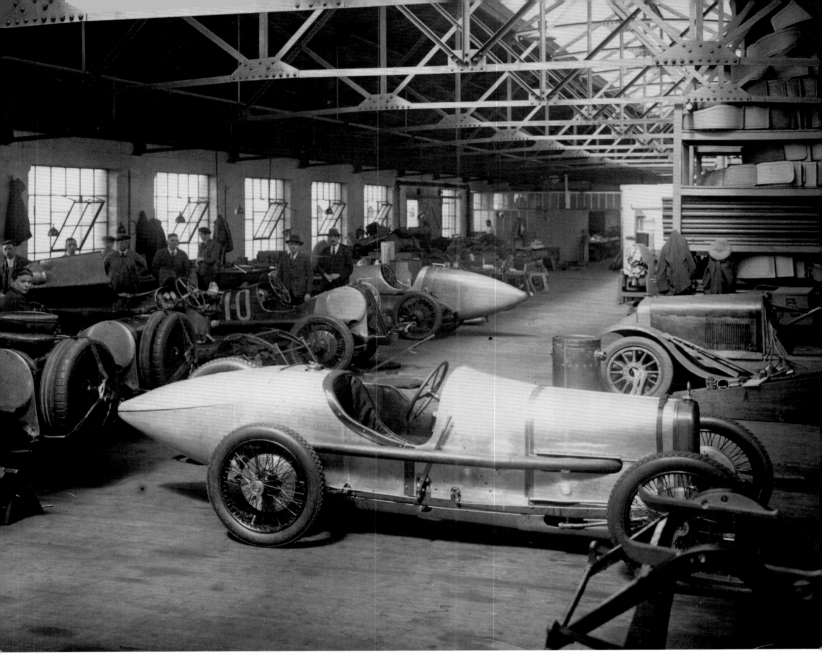

Sunbeam Racing and Experimental Department, March 1922. In left foreground are four straight-eight, 3-litre cars being rebuilt for the Tourist Trophy Race. Behind them is the pointed tail of the 18-litre, 350 hp racing car and in the centre one of the new 2-litre, four-cylinder Grand Prix cars. Louis Coatalen and Jack Irving are standing in front of the 350 hp car.

Talbot introduced a little 8 hp, 970 cc car with quarter elliptic springs described in the press as 'really outstanding'. It had been designed and developed in France although it was to be produced at Barlby Road. In France, Talbot-Darracq had introduced a 12 hp car from which the 14 hp Sunbeam was developed.[17]

The report of the Directors of STD Motors showed a net profit of £151,382 for the year which they felt was 'satisfactory having regard to the conditions of the past year'.

---

17. The integral aluminium cylinder block and crankcase of the 14 hp Sunbeam were soon replaced with a cast iron block on an aluminium crankcase. The London Talbot version of the car was discontinued by early 1922.

Chapter Five

Capt. Jack Irving and Louis Coatalen with a newly built 1922 Strasbourg Grand Prix Sunbeam.

1922: From the Isle of Man to Sicily

In March 1922 Louis Coatalen invited a journalist from *The Autocar* to visit Sunbeam's experimental and racing department at Wolverhampton to see the cars being prepared for the forthcoming season. They were working on more than a dozen racers; the 350 hp twelve-cylinder car had been overhauled and repainted, four of the 3-litre, straight-eights designed for the previous year's Grand Prix were being rebuilt for the Isle of Man Tourist trophy race, three new 2-litre, four-cylinder cars were being constructed for the Strasbourg Grand Prix, there were three of the 1½-litre Talbot-Darracqs (potentially masquerading as Sunbeams), and two special, dark blue, Q cars which had 1921 GP chassis fitted with 1919 4.9-litre, six-cylinder 'Indianapolis' engines. One of these two was a single-seater built with the aim of setting up a record of 100 mph for twenty-four hours, while the other was fitted with a two-seater 'semi-streamlined body'. It is notable that, apart from the cars for the Grand Prix, all of them were developments of existing cars, so expense was contained.

Despite the fact that the fashion for eight-cylinder engines was temporarily waning, Coatalen was keen to draw attention to their advantages, claiming high revs, better balance and improved torque as a result of the lighter reciprocating parts. He also enthused about the effectiveness of the four-wheel braking system on the Grand Prix cars which incorporated a servo motor. He remarked, with a twinkle in his

A New Beginning? 1919–23

Kenelm Lee Guinness spins the wheels of the 350 hp, 18-litre Sunbeam as he sets off to create standing start World's Records for the half-mile, the kilometre, and the mile on 17 May 1922. Standing behind the car are Sunbeam mechanics Jack Ridley and Bill Perkins.

eye, 'when the car comes up to a corner, unless the mechanic holds tight, he will be left in front'.

1922 was the year when the real potential of the 350 hp car emerged after its early tribulations. In April, Coatalen permitted the journalist S.C.H. Davis to try it at Brooklands and he described in glowing terms what an exhilarating experience this was. Soon after, Jean Chassagne won a race there with it lapping at 114 mph, but it was Kenelm Lee Guinness who got the best out of it at Brooklands by setting up a number of World Records in May. As the records required two-way runs this also involved going the 'wrong way' round the track at 140 mph, which *The Motor* described as a 'breathlessly thrilling' spectacle. Some idea of the power of the vehicle can be gathered from the description of the standing start mile record set at 96.63 mph; the rear wheels spun furiously when the clutch was engaged, leaving black parallel lines for 30 yards and they spun again as second gear and third gears were selected. For the flying mile a mean speed of 129.17 mph (133.75 mph for the kilometre) was established while the quickest flying half-mile recorded was at 140.51 mph.

In early June the Sunbeam and the Talbot-Darracq teams arrived on the Isle of Man to prepare for the Tourist Trophy races where the Sunbeams were to compete in the 3-litre race and the Talbot-Darracqs in the 1½-litre race, to be held concurrently on 22 June. A team of 1½-litre Sunbeams, which had been mooted in addition to the Talbot-Darracqs, never materialised. As in 1914, Coatalen made his headquarters at the Fort Anne Hotel in Douglas

Chapter Five

where there was a pleasant secluded courtyard at the rear with good sheds for the Sunbeams, although the Talbot-Darracqs had to make do with smaller sheds on the street frontage which constantly attracted a crowd of curious onlookers. Once more, practice took place for a fortnight between 4.30am and 7.00am, and again both the Guinness brothers were to drive but they were joined by Segrave, Chassagne, Moriceau and Divo. Sadly, the Sunbeam of Kenelm Lee Guinness, who had won in 1914, was unable to start due to clutch problems, so it was Segrave who led from the start in pouring rain and a sea of mud to put up the fastest lap until he had to retire with magneto failure. Chassagne on the remaining Sunbeam who had been running in second place then stepped up and, thanks to the four-wheel brakes, was able to hold off the challenge of Frank Clement aboard the first of the Bentleys, to win the race in 5 hours, 24 minutes and 50 seconds at an average of 55.78 mph. Meanwhile, in the shorter 1,500 cc race, the Talbot-Darracqs were once again dominant with Sir Algernon Lee Guinness winning at an average speed of 53.3 mph, followed in second place by Albert Divo. For a time it looked as though it might be another 1-2-3 finish for the Talbot-Darracqs but Jules

Top: The STD team for the 1922 Isle of Man Tourist Trophy Race. In the front row are the 1½-litre Talbot-Darracqs driven by A. Divo, A.L. Guinness and J. Moriceau. Behind are the 3-litre Sunbeams of K.L. Guinness and H.O.D. Segrave. The car of the eventual winner, Jean Chassagne, is absent.

Centre: Jean Chassagne, winner of the very wet 1922 Isle of Man Tourist Trophy race for 3-litre cars at the wheel of his eight-cylinder Sunbeam.

Bottom: Sir Algernon Lee Guinness, winner of the 1922 1½-litre, Tourist Trophy Race with the four-cylinder Talbot-Darracq.

Moriceau had approached a corner too fast and the car overturned, putting him out of the race.

The Grand Prix de l'Automobile Club de France was held on 16 July at Strasbourg. At the previous year's Grand Prix eleven of the thirteen starters had used straight-eight engines, a fashion that owed a great deal to the eight-cylinder Ballot racing cars designed for the 1919 Indianapolis race by Ernest Henry. However, even before the 1921 Grand Prix it seems as though that fashion was on the wane. In an article published in *The Autocar* on 7 May 1921, W.F. Bradley wrote that 'engineers look upon the straight-eight as mechanical progress, while drivers are inclined to consider it a passing fashion'. With the capacity limit reduced from 3 litres to 2 litres one might have expected the latest cars to have six-cylinder engines but Ernest Henry's view, recorded by Bradley, was that such engines produced vibrations and difficult carburation problems. Drivers such as René Thomas and Jean Chassagne favoured a return to four cylinders.

Ernest Henry was, of course, the designer of the pre-war Peugeots which Coatalen had diligently copied in 1914. Coatalen had developed those designs throughout the war but at the end of 1921 Henry himself was recruited to take technical control of the Talbot-Darracq racing department in Suresnes and to design the 1922 Grand Prix Sunbeam. The eight-cylinder engines used during the previous year were replaced by a new four-cylinder engine that bore many of Henry's hallmarks but a distinctive novelty was the asymmetric design of the overhead valve gear to permit larger inlet valves. Despite optimism before the race that the cars were highly competitive, the appearance of the much faster Fiat team, equipped with six-cylinder engines, perturbed Coatalen and his drivers. Axle ratios were lowered, as it was

Top: The 1922 Sunbeam Grand Prix team at Strasbourg. Left to right: Jean Chassagne and Robert Laly, Kenelm Lee Guinness and Albert Divo, Henry Segrave and Jules Moriceau. Louis Coatalen stands behind the cars of Guinness and Segrave, sporting a bow tie. Bill Perkins in cap.

Bottom: Louis Coatalen looks very happy to allow his friend, the journalist Charles Faroux, a test-drive in one of the 1922 Grand Prix Sunbeams at Strasbourg.

Above: The Talbot-Darracq team lined up for the start of the 1922 200 Mile Race at Brooklands. Chassagne no. 5 crashed after a tyre burst. Guinness no. 6 won the race and put up the fastest lap at 95.78 mph. Segrave no. 7 finished third.

Left: Louis Coatalen supervises Jean Chassagne and Paul Dutoit, making last-minute adjustments to their 200 Mile Race Talbot-Darracq.

believed that the engines were not achieving their peak rpm, but it was subsequently found that the rev-counters were under-reading and so the engines were being over-revved in the race. The result was that all three of the Sunbeams retired with broken inlet valve stems. Ernest Henry blamed the failure on changes to his design that Louis Coatalen had insisted on making and they parted company on bad terms.

After this setback, the next challenge was the 200 Mile Race at Brooklands on 20 August for which the little Talbot-Darracqs were re-fitted with the streamlined bodies and the engines were mounted higher in the frames. Segrave led from the start but experienced misfiring problems halfway through and finished third. Chassagne's car misfired from the start but eventually got going properly until a tyre burst and the car disappeared over the top of the banking, fortunately without serious injury to driver or mechanic. It was left to Kenelm Lee Guinness to save the honour of the firm and to win the race at the same speed (88 mph) as the previous year. After that it was back to Le Mans in September for the

A New Beginning? 1919–23

Coupe des Voiturettes where the little Talbot-Darracqs, which were frequently likened to roller skates, dominated again, taking the first three places with Guinness first, Divo second and Segrave third. Although the opposition was weak Guinness kept up a cracking pace covering the 375 miles at an average of 72 mph on the triangular road circuit.

This would have been an appropriate ending to the racing season. Although the Sunbeams had failed in the Grand Prix, they could at least point to their victory in the Tourist Trophy race and the World Land Speed Record, while the Talbot-Darracqs had an unbroken record of having won every long-distance race for which they had been entered. However, somebody came up with the idea of a Mediterranean cruise!

Kenelm Lee Guinness had purchased an ex-tramp steamer which had been fitted with cabins and the hold was converted to accommodate several cars. KLG generously put this 'yacht' at the disposal of the STD racing team free of charge. Kenelm and Algernon Guinness set sail in the *Ocean Rover* on 15 October from Southampton with H.O.D. Segrave and the mechanics as passengers, carrying two Sunbeam and two Talbot-Darracq racing cars with the aim of taking part in a race in Spain and then the Coppa Florio in Sicily. Both Guinness brothers were accomplished navigators but the storms they encountered crossing the Bay of Biscay were such that they had to seek shelter at Coruna before sailing on round to Barcelona. There they were joined by Jean Chassagne with the third Talbot-Darracq and Louis Coatalen who had driven down from Paris. The Penya Rhin Grand Prix held at Villafranca on 5 November was the last voiturette race of the year and it was to be the final race for those Talbot-Darracqs. It came close to being the race that would break their previously unbeaten record, as the Chiribiris were faster and even the Aston

Top: The Talbot-Darracq Team at Le Mans 1922 for the Coupe des Voiturettes race. Guinness no. 2 won the race, Divo no. 5 was second, Segrave no. 4 was third.

Bottom: Segrave at Barcelona 1922 being push-started by Jules Moriceau for the Penya Rhin Grand Prix. Segrave finished fourth but the race was won by his Talbot-Darracq team mate K.L. Guinness.

Martins were going as fast. However, various mishaps befell the opposition and Kenelm Lee Guinness was able to come through to win. The rest of the team, for once, were not just behind as Segrave finished only in fourth place after an inlet valve broke and the car caught fire several times (Coatalen was unsympathetic and recommended, if the car caught fire again, that Segrave should get out and step in a puddle). Chassagne had retired with valve problems early on. Guinness's car was driven back to Barcelona and left on display in the agent's showroom while its place in the hold of the ship was taken by Louis Coatalen's Sunbeam touring car.

The *Ocean Rover* took four days to sail from Barcelona to Palermo, encountering more very rough weather on the way that forced them to take shelter by Sardinia, but they still had a week to get prepared for the Coppa Florio race in which the two Sunbeams were to take part. These were fitted with the powerful 4.9-litre six-cylinder engines and to be driven by Segrave and Chassagne. It seems the team had rather underestimated the difficulty of the 67-mile circuit which climbed to over 3,500 ft and had around 1,600 corners. On a practice lap Segrave, delayed by a mechanical problem, was caught out by the suddenness with which night fell and he and his mechanic Moriceau had to sleep in a barn. When they did not return, Coatalen sent out a search party at 8.00pm but the searchers returned empty-handed to the ship at 1.30am. Segrave recorded that they persuaded Coatalen to drive round the circuit with them in a touring car so that he could see for himself the difficulty the drivers were having to cope with. They stopped high up in the mountains for a picnic lunch. 'Nobody had said very much, but Mr Coatalen had said absolutely nothing during the whole run. Afterwards, when the silence had become almost oppressive, he suddenly looked round at us and said in a melancholy voice: *"Quel spectacle de désolation."* And that exactly describes the Madonie circuit.'[18]

During the race Segrave came across another competitor's car which had overturned, injuring the driver and killing the mechanic. He and

18. H.O.D. Segrave, 'Four Years of International Motor Racing', *The Motor*, 21 October 1924, p. 579.

Top: Kenelm Lee Guinness's yacht *Ocean Rover* in the port of Termini Immerese near Palermo in Sicily, having safely transported the Sunbeam racing cars from England via Spain.

Bottom: Segrave's 4.9-litre, six-cylinder Sunbeam on the quay having been unloaded from *Ocean Rover*, which is just behind it.

Moriceau stopped and helped them to a place of safety. As they raced on they encountered various other problems, such as a leaking radiator and oiling plugs but managed to finish in second place after eight-and-a-quarter hours of motoring. Chassagne, to whom Coatalen had waved the red, go-faster flag on the first two laps, was less fortunate as an oil pipe broke which resulted in him losing all his oil and he had to beg locally produced olive oil up in the mountains in order to drive gently back to the finish. By the time he arrived the officials had packed up so he was not placed.

After the race *Ocean Rover*, with her cargo of cars back on board, was used for a little more relaxed cruising up the coast of Italy to arrive at Monte Carlo a week later. There, Coatalen's car was put ashore to be driven back to Paris by two of the mechanics, Bill Perkins and Paul Dutoit. From Paris they made their way back to Wolverhampton by train and ferry. Perkins' list of faults on the car at the end of this trip included the engine pulling badly, worn-out brakes and broken shock absorber brackets. The *Ocean Rover* arrived back in England before Christmas, but the weather had been so bad that yachting cruises were henceforth a taboo subject amongst those who had taken part.

Jean Chassagne and Paul Dutoit working on the former's car for the 1922 Coppa Florio Race in Sicily.

Chapter Five

Chassagne at the start of the Coppa Florio Race. Segrave, in white helmet, stands beside the car and Louis Coatalen is on the right in cap and overcoat.

Running this extensive racing programme was no doubt time-consuming and we know that Coatalen was always on the front line at each competition supervising operations and signalling his drivers to go faster. He must have covered many thousands of miles between the different factories and the geographically scattered race circuits. But he had other technical responsibilities within the group, which were further complicated by William Iliff being out of action during May and June with appendicitis, and the engineering employers' lockout around the same period. Despite the lockout, through which employers attempted to reverse gains such as control over overtime made by the Amalgamated Engineering Union during the war, the Sunbeam works were reported to have managed to produce around fifty cars per week.

It is through articles published under Coatalen's name in *The Motor* and *The Autocar* during 1922 that we glean details of the design and research work being undertaken. Evidently the articles were intended to promote STD products and justify his decisions, but they also give an insight into his view of the role of the designer. They confirm his view that 'the production of better touring cars is largely dependent upon the lessons learnt in racing practice.... The racing car of today is proverbially the touring car of tomorrow.'[19] Before World War I Coatalen had developed side-valve engines to produce greater output than had ever been obtained from that type of engine, but racing had then taught him that overhead

---

19. Louis Coatalen, 'With & Without Differential', *The Autocar*, 6 May 1922.

valves were even better. He had experimented during 1920 and 1921 with a shaft-driven single overhead camshaft but it proved to be unacceptably noisy. As a compromise for the production cars, Herbert Stevens came up with a proposal to adapt the existing side-valve crankcases to pushrod operated overhead valves, thereby giving the sales force an overhead valve engine to sell and enabling the existing stock of crankshafts and crankcases to be used up. Thus it was that the 16 hp and 24 hp models 'suddenly' had overhead valve engines for 1922. In an article entitled 'Overhead Valves – Advantages and Disadvantages' in *The Motor*, Coatalen stressed that 'the overwhelming superiority of the overhead valve system is very definitely decided' and then summarised those advantages as: greater output, decreased heat losses, better consumption, better gas-flow, accessibility, better lubrication, and less weight. The conversion of a Sunbeam engine had produced a 25 per cent increase in power. He concluded that 'the question as to whether overhead camshaft or push rods and rockers should be used is determined by the class of the engine.... For very high speed engines the overhead camshaft is essential.'[20] The article provoked a response from T.D. Wishart, the Crossley designer, who described himself as a wholehearted believer in the side-valve engine for its simplicity, fewer moving parts and silence. Coatalen and Wishart were able to keep the subject simmering in the columns of *The Motor* for nearly two months.

Similarly, it was Coatalen who reacted to an article in *The Autocar* by Granville Bradshaw in favour of differentials in the back axles of light cars. In view of the fact Talbot had recently introduced the 8 hp model without a differential, Coatalen could not let this pass and pointed to the success of his various differential-less racing cars. He listed fourteen points on which he differed and condescendingly, with a typical Coatalen twinkle, wrote, 'I will give Mr Bradshaw the satisfaction of knowing that if I were to design a light car for gyrating around flower beds I should certainly incorporate a differential in the rear axle. But I conceive that light cars are not commonly used for this amiable and innocent purpose.' A couple of weeks later Bradshaw responded at length, saying that, as most of Coatalen's illustrations were based on racing, they 'constituted no answer of any sort' and suggested the problems of controlling the 350 hp Sunbeam experienced by different drivers were due to the lack of differential. This opened the way for Coatalen to disagree and propound again his belief that 'the production of better touring cars is largely dependent upon lessons learnt in racing practice'. He pointed out that overhead valves were the product of racing practice, that the stronger, lighter nickel-chrome crankshafts existed because they had been demanded by racing, and that 'dry sump lubrication, in conjunction with a forced system, was first introduced by myself as a result of the 12-hour record at Brooklands'. He revealed that when Hawker had gone off the track at Brooklands with the 350 hp Sunbeam it had been fitted with a differential. It was later after further experiments that the decision had been made to dispense with the differential.

20. *The Motor*, 5 April 1922.

Chapter Five

The 8-18 Talbot powered by a 970 cc overhead valve engine, with ¼ elliptic springs all round and no differential was introduced for 1922. The actress Gladys Cooper is setting off for a round of golf.

He explained that, 'the solid axle would avoid a torque effect which has proved distinctly unwelcome. When the 12-cylinder is running at full throttle the torque of the engine is so great that the load on the off-side wheel is greatly reduced with the result that, when a differential is used, the wheel spins.'

The design of the small 8 hp Talbot took Coatalen into an area in which he had little previous experience but he maintained the car was so diminutive that it had no need of a differential. What was vital was the ratio of sprung to unsprung weight and a lighter rear axle was thus more important. In a later four-page article he described the extensive testing that had been carried out to develop a satisfactory suspension system. Using a system of lights attached to the rear hub and the body, the car was driven past a camera at night, numerous times, to create a record of the effects of different loads, different springs, spring lubrication, shock absorber settings, etc. Coatalen admitted that he was one of those who had in the past been guilty of neglecting suspension design in favour of engine development, but he ascribed this largely to public demand and revealed his faith in market forces. He wrote:

> the laws of supply and demand are inexorable. It is in fact the first duty of the automobile manufacturer to

> attain commercial soundness … the true criterion of good design is the dividend that it ultimately pays to those who have supported it financially…. The public has by no means been so interested in suspension developments … car designers have taken the line of least resistance. It is not, and never can be their part to educate the public…. After all it is the buying public which really decides the form that the newest type of motor car shall take, the designer's function is to put that decision into practice.

The prevailing economic climate meant that the demand was for less expensive cars, which in turn meant lighter cars at a time when road surfaces were getting worse. Lighter cars going faster and variations in load that altered significantly the ratio of sprung to unsprung weight presented designers with new challenges. He admitted that he did not have the solution to these problems, saying, 'incidentally I question whether any one man is capable of finding the solution to the problem – the only thing to do is to experiment and to find out as much as one possibly can in the hope that eventually one may get on the right track'.[21]

He wrote a further article which was published the following month, giving results of systematic experiments using similar recording techniques carried out on a car with no shock absorbers, then fitted with frictional dampers, and then with elastic shackles and different combinations of both. This time a lamp was also fixed to the rim of the rear wheel to trace a trochoid curve which revealed unexpected distortions.[22]

1923: The Grand Prix Victory

Ernest Henry was not the only disgruntled designer at the Strasbourg Grand Prix in 1922. In the Fiat pits was a young man, Vincenzo Bertarione, who had been involved with designing their victorious car but who felt that he had been inadequately rewarded for his efforts. He talked about his frustration to Edmond Moglia, another Italian then working in the Talbot-Darracq design office, who passed the information on to Louis Coatalen. Coatalen was a firm believer in recruiting the best talent and soon made Bertarione an offer to join the design team in Paris. At the beginning of August 1922, just three weeks after the Grand Prix, Bertarione signed a two-year employment contract to be head of the Talbot-Darracq racing design office starting in November. When he started work he brought with him his friend and assistant Walter Becchia.[23] Within six months Sunbeam had a new six-cylinder, 2-litre Grand Prix car that was very different to its four-cylinder predecessor but very similar to the Fiat. It had twin overhead camshafts driven from the rear and only two valves per cylinder compared with the four valves of its predecessor. The other notable difference was that in place of a cast iron cylinder block the new car had lighter welded-on water jackets around forged

---

21. Louis Coatalen, 'The Problem of Suspension', *The Autocar*, 22 September 1922, p. 537.
22. Louis Coatalen, 'A Study of Suspension', *The Autocar*, 20 October 1922, p. 757.
23. Sébastien Faurès, *Fiat en Grand Prix*, ETAI, 2009.

Chapter Five

Top left: Louis Coatalen's two sons, Hervé and Jean, in one of the 1923 Sunbeam Grand Prix cars a few days before the race.

Top right: This snapshot was taken by Hervé Coatalen, aged ten, in the backyard of the Boeuf Couronné inn at Neuvy-le-Roi, near Tours. Segrave is at the wheel of his car while his mechanic, Paul Dutoit, in white overalls walks alongside.

Bottom: The atmosphere was relaxed as the mechanics fool around with a patient donkey at the team's base in Neuvy-le-Roi. Tommy Harrison and Jack Ridley continue to work on Guinness' car despite the distraction.

steel cylinders. These were all features to be found on the Fiat but the Sunbeam engine also incorporated distinctive differences to its Fiat inspiration, such as gear-driven instead of shaft-driven camshafts, and asymmetric valve angles to accommodate larger inlet valves. Albert Divo picked up one of the new cars in Wolverhampton and tested it, along with Segrave, at Brooklands in May 1923; Divo was soon to be seen trying it out on the Grand Prix circuit near Tours before the roads were closed to such activity. As the date for the race approached, Guinness and Segrave also drove their cars from England to the Talbot-Darracq works in Suresnes for final changes before the Sunbeam team installed themselves in a 'half farm, half country inn', Le Boeuf Couronné at Neuvy-le-Roi, in the Touraine countryside. 'Under the indifferent eye of a passive mule and an indolent pig the mechanics could work in the long stone-floored hall transformed into a workshop, to the occasional accompaniment of a wheezy foxtrot played by a mechanical piano.'[24] This was in marked contrast to the Fiat team who were installed in the altogether grander surroundings of Château Poillé.

Once again, the confidence of the Sunbeam team was undermined by the appearance of the new Fiat racing cars. Bucking the trend for the second year, the Fiats had gone for eight-cylinder engines which were also fitted with a previously unseen development in Grand Prix racing – a supercharger. It was a rotary blower

The six-cylinder, 2-litre engine for the 1923 Sunbeam Grand Prix cars.

driven directly off the crankshaft and delivering compressed air to the carburettor. Segrave's fastest lap in the official practice session on the Sunbeam was at 81 mph, while Bordino in his Fiat managed 85.6 mph. Coatalen and his team had obtained 108 hp from their engine which was substantially more than the previous year's Fiat (92 hp) but this new eight-cylinder supercharged Fiat produced 130 hp. In the race Bordino dominated the first eight laps as he demonstrated the potential of the Fiats. He set up the fastest lap of the race at 88.05 mph but then had to retire. Kenelm Lee Guinness, who had been in second place showing his skill as a driver by averaging 82 mph, then took the lead for four laps until his clutch started to slip and he dropped back down the field. Giaccone

24. *The Autocar*, 29 June 1923.

Chapter Five

and Salamano in the two remaining Fiats were then first and second until Giaccone had to retire with a broken exhaust valve caused by the grit being sucked through by the supercharger. After a slow pit stop by Salamano, Divo's Sunbeam held the lead for a while but his own pit stop turned out to be a much worse disaster as the fuel cap jammed and from then on, he had to stop on every lap to top up the small reserve tank under the scuttle. Segrave in the third Sunbeam had deliberately set out to drive a conservative race lapping at a steady 75 mph. Despite also having problems with a slipping clutch, his car had not been overstretched so that, as the other competitors dropped out, he found himself working his way up the scoreboard and his car running better and better. With just a couple of laps to go, Salamano, who had been comfortably in the lead on the remaining Fiat, came to a halt with engine failure and thus Segrave led the race. He completed the 496-mile race in 6 hours and 35 minutes at an average speed of 75.3 mph, followed home in second place by Divo, with Guinness coming fourth behind Friedrich's Bugatti. Of the seventeen starters only five finished, so the Sunbeams were the only complete team still running and were able to go home victorious. This was the first British car ever to win the French Grand Prix, and it was a feat that was not to be repeated for thirty-seven years.[25] In the Sunbeam box in the grandstand, the race had been witnessed by Arthur Huntley Walker and Owen Clegg (both directors of STD Motors), Mrs Owen Clegg and Mrs Louis Coatalen. They had lived through an emotional roller-coaster ride as the team's position had fluctuated and they experienced hope, desperation and finally joy. Only one person remained calm and imperturbable throughout – Iris Coatalen.

---

25. In 1955 a Connaught won the Syracuse Grand Prix but the next British car after Sunbeam to win the French Grand Prix was the Cooper Climax driven by Jack Brabham in 1960.

Top: The start of the 1923 Grand Prix at Tours on 2 July. Bordino (Fiat) and Thomas (Delage) are already out of sight. No. 2 is K.L. Guinness (Sunbeam), followed by Guyot (Rolland Pilain), then Friedrich (Bugatti) and Duray (Voisin). No. 9 is Giaccone (Fiat) and no. 7 is Divo (Sunbeam).

Bottom: Henry Segrave, the winner of the Grand Prix, passing the pits in his 2-litre six-cylinder Sunbeam. He took the lead just two laps from the end when Salamano's supercharged Fiat retired. He won at an average speed of 75.3 mph.

The heroes return. After winning the Grand Prix in France, Henry Segrave and Louis Coatalen are given an enthusiastic welcome by the workforce back at the Sunbeam factory in Wolverhampton.

By coincidence it was announced on the same day that Louis Coatalen had been nominated Chevalier of the Légion d'Honneur in recognition of his services during the war as a designer and producer of aviation engines for the Allies. The French Government thus went some way to acknowledging Coatalen's contribution which the British Government had notably failed to do. He had plenty of causes for celebration in July 1923.

When the team was back in England the celebrations continued with a luncheon given by the RAC at which James Todd, Chairman of STD Motors, and Louis Coatalen were the guests of honour. The drivers and mechanics were presented with commemorative medals by Sir Arthur Stanley and congratulatory speeches were made to which the drivers responded. Unusually, Louis Coatalen was loquacious in his speech, paying tribute not just to the members of his team but also to the Fiat team who had tried to make the pace too hot. He insisted that 'we have learned lessons

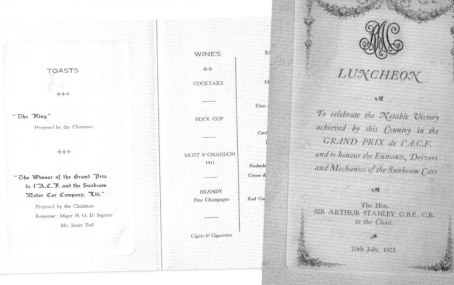

Menu for the celebratory luncheon held by the RAC on 20 July 1923 to honour the entrants, drivers and mechanics of the Sunbeam Cars.

Chapter Five

which we shall be able to embody permanently in our touring car practice' and he concluded by expressing the hope that in 1924 a British Grand Prix would be held and suggested that Brooklands could be suitably modified with an artificial corner in the finishing straight.

Meanwhile, Bertarione had also produced a 1½-litre, four-cylinder version of his racing engine along with a matching chassis so that a new team of Talbot-Darracq racing cars could be built. Although they were ready by early August and producing 70 hp at 5,000 rpm, the knowledge that Fiat would be supercharging their cars for this class of racing as well, upset the plans. Coatalen was not at all keen to confront the Fiats directly, arguing that it was not fair play to make use of supercharging, as it effectively increased the capacity of the engines. Although the Sunbeams had won at Tours, the potential of the Fiats had been clearly signalled. In the words of Laurence Pomeroy, 'the 1923 Grand Prix is perfect proof that the first concept of superior principle is always defeated by the perfected example of established practice'.[26] But it was only a matter of time before the new technology was made to work reliably.

The first race outing for the Talbot-Darracqs was at the beginning of September in the Light Car Grand Prix at Boulogne where the Fiats were not entered. Segrave won comfortably at an average of 65.35 mph, although both his team mates, Guinness and Divo, had to retire. Three weeks later the Talbot-Darracq team appeared at Le Mans for the Coupe des Voiturettes, this time with only French drivers – Divo, Bourlier and Moriceau. Here there was even less competition as there were just two other entries so Divo cruised comfortably to victory at 71.5 mph followed by Moriceau in second place after Bourlier retired.

The 200 Mile Race at Brooklands in October should have been the next event for these little 1,500 cc racers but they were not entered as it was

Posing for a light-hearted photograph in the Sunbeam Experimental Department in 1923. The vehicle is a Sunbeam Mabley manufactured at Wolverhampton between 1901 and 1903, powered by a single-cylinder De Dion engine. Left to right: C.B. Kay, L. Coatalen, unknown, K.L. Guinness, T. Harrison, P. Dutoit, J. Power (apprentice), H.O.D. Segrave, H.C. Stevens, J. Irving. It was restored and took part in a number of London to Brighton Runs in the late 1920s.

---

26. Laurence Pomeroy, *The Evolution of the Racing Car*, William Kimber, 1966, p. 135.

Grand Prix de Boulogne 1923, the first outing of the new Talbot-Darracqs. Segrave seen here, with Paul Dutoit as mechanic, won and put up the fastest lap.

known that the Fiats had been entered. The 2-litre, eight-cylinder Fiats had in the meantime won the European Grand Prix at Monza in convincing style demonstrating they had learned useful lessons at Tours, so the 1,500 cc versions (fitted with Roots type blowers in place of the Wittigs that had given trouble in the Grand Prix), to be driven by Salamano and Campbell, were even more likely to be a real threat to Talbot-Darracq supremacy. In the event both Fiats, having put up the fastest laps, retired when their engines blew up as a result of being pushed too hard, too soon, and the race

was won by an Alvis. But the Talbot-Darracq team was already on its way to Spain, arguing that continental commitments took precedence, so the confrontation was avoided.

Both the Talbot-Darracqs and the Sunbeams must have used more conventional means of travelling south than they did in the previous year, as Kenelm Lee Guinness was not part of the team. At the Penya Rhin Grand Prix held on the Villafranca street circuit some 250 miles inland from Barcelona, Divo dominated with his Talbot-Darracq and Dario Resta, back from America, in the second car was third but some

Chapter Five

Left: Louis Coatalen, Dario Resta and Albert Divo at the opening event of the new Sitges track near Barcelona in October 1923. Divo won the 2-litre class at 97 mph with the six-cylinder Grand Prix Sunbeam.

Right: Dario Resta no. 9 and Albert Divo no. 14 on their 1500 cc, four-cylinder, 8 valve, Talbot-Darracqs for the III Gran Premio do Penya Rhin held in October 1923. Divo won, Resta finished third behind Zborowski's Aston Martin. Louis Coatalen stands behind Resta's car.

35 minutes behind the leader. Moriceau was unable to start as a spare part that was required was held up in customs. A week later Divo and Resta, driving Grand Prix Sunbeams, took part in the inaugural 400 km race for 2-litre cars on the newly constructed, steeply banked, race track at Sitges, 25 miles down the coast from Barcelona. Divo won at an average of 97 mph, being just 50 seconds ahead of Zborowski's Miller which had led until a tyre burst two laps from the finish; it was pointed out that this was faster than the speed of the winners of both the European Grand Prix and the Indianapolis race. Resta had retired after 150 laps. A few days later they were out again with the Talbot-Darracqs on the same circuit for the even longer Voiturette Grand Prix. This time it was Resta who won at 85 mph, although Divo put up the fastest lap and led for most of the race until he slowed slightly, allowing Resta to take the flag just 10 seconds ahead. None of the drivers enjoyed racing on this brand-new circuit, being of the opinion that the banking had been incorrectly designed.

If the racing department could point to two Grand Prix victories (Sunbeam) and four Voiturette race victories (Talbot-Darracq) in 1923, there was no spectacular change to report in the rest of the organisation. Co-operation between Sunbeam and Talbot was reinforced by the appointment of L.V. Cozens, who was sales manager for Sunbeam, taking on the additional role of joint sales manager of Clement Talbot with A.D. Makins. Sadly, Walter Coombes, Sunbeam's long serving publicity manager died unexpectedly but not before he had made it known that Sunbeam had made a model 24-60, six-cylinder Sunbeam for the Queen's Dolls House at Windsor Castle. Trade was difficult and prices of all Sunbeam models had to be reduced, as was that of the 10-23 Talbot. At the end of the year, shortly before Christmas, the Wolverhampton works was visited by the Prime Minister of New Zealand but Coatalen was not present; William Iliff and C.B. Kay showed him round.

In the French Alps, Coatalen severely tested one of the new six-cylinder Sunbeams, equipped with a brake servo, and two Talbot-Darracqs successfully completed the 2,500-mile Tour de France without losing points. Business appeared to be good at the Darracq works but one problem was finding sufficient skilled labour, so women had to be employed in fairly large numbers. However, Owen Clegg was reported to be optimistic as the depressed value of the French franc meant the absence of foreign competition and orders continued to flow in, particularly for the light cars that were encroaching into the big car market.

Coatalen continued to plead in favour of racing with an article in *The Autocar* entitled 'How Racing Influences Touring Cars', which concluded, 'very great benefit is obtained from the study of racing engine performances and I am firmly of the opinion that money spent by manufacturers on racing work will result eventually in almost startling improvements in touring car units'.[27] Sir Herbert Austin wrote to admit that 'racing has undoubtedly hastened our knowledge and experience ... and given data in weeks that would have taken years by any other means' but, he continued, 'it has always seemed to me that building of special or freak cars is a mistake. I have never attempted to go outside equipping a standard car with some small change' that should be capable of being sold in large quantities. It was left to Warwick Wright, leading STD sales agent, to respond, 'if you want to improve the breed ... you must get that new blood from special racing designs'.

During 1923 STD motor racing efforts had been cut back and were more targeted in comparison to the previous year. They were noticeably absent from racing at Brooklands, apart from an autumn meeting where mechanic Bill Perkins drove one of the eight-cylinder cars. The 350 hp Sunbeam was sold to Malcolm Campbell (along with one of the 4.9 litre Coppa Florio cars) who used it to push the Land Speed record even higher.

At the end of the year Coatalen revealed that it was his intention to enter three British-built, Paris-designed Sunbeams for the 1924 Grand Prix to be held at Lyon which would be driven by Guinness, Segrave and Resta. 'Mr Coatalen admits that the new Sunbeam racers will have a supercharger device, for it is impossible to compete on an equal basis under the present rules without such an addition ... any objections he had raised were based on its use under a piston displacement rule – not to the principle of the appliance.'

---

27. *The Autocar*, 1 June 1923, p. 939.

Chapter Five

Private Life

It is clear that an affair between Iris van Raalte and Louis Coatalen commenced during World War I. In 1917 it was well underway and the following year Louis separated from his wife Olive, moving to live in a flat in Jermyn Street, London with Iris who had left Noel van Raalte when her second daughter (Charmian, b. 1916) was less than a year old.

Iris Enid Florence Graham (1891–1976) was certainly an unusual woman and someone who seems to have been determined continuously to challenge convention. She was the only child of James Graham (second son of Sir Frederick Graham, Baronet of Netherby) and his wife Florence Rose (née Carter-Wood) whom he had met while travelling in India. Iris was therefore 'a niece of Lady Cynthia Graham, of the Duke of Montrose, of the Earl of Harrington and the Earl of Verulam, and a cousin of the Duke of Leinster'.[1] In stark contrast to this aristocratic background, as a young girl brought up in Cumberland, she learned to shoot, poach salmon and trout, and breed rats for medical research on the family estate which comprised thirty-

This portrait of Iris by the fashionable society painter David Jagger was displayed at the Royal Academy in May 1918 while she was still Mrs van Raalte. It was stolen from the home of her daughter Marjolie in the South of France in 1994.

nine hill farms managed by her father. She married the wealthy Noel van Raalte in November 1912 and their first daughter was born in 1913. According to Marjolie Coatalen, Iris chose the name Gonda for this daughter in order to upset her parents-in-law.[2] A second daughter, Charmian, was born in 1916 but (again according to

---

1. *Tatler*, 21 February 1923.
2. Marjolie Coatalen, draft autobiography typescript, chapter 2, p. 9. Iris was no doubt inspired by the musical *The Girl on the Train*, which had been produced at the Vaudeville Theatre, London, in 1910. The two leading characters were Karel van Raalte and Gonda van der Loo.
The website www.theguidetomusicaltheatre.com gives the following synopsis of the plot: 'Gonda van der Loo, a young actress travelling on a train in Holland at night, is unable to secure a berth. Karel van Raalte, a young gentleman, generously offers his compartment to her. The two, however, become locked in the compartment. Their cries and knocks are unheard and they are forced to spend the night together. Van Raalte's wife learns of the incident and jealously brings divorce proceedings. After many complications and much time spent in the divorce court, van Raalte and his wife are reunited and the judge and the actress find romance together.'

Marjolie) the real father of this child was Louis Coatalen.³

When Louis and Iris set up home together they seem to have been happy to turn over a page and start with a blank sheet. Coatalen's two sons, Hervé and Jean, were left with Olive whilst the two van Raalte girls, Gonda and Charmian, were left with Noel. The latter was certainly unhappy about the break-up of his marriage, as it was reported that when he found out what was happening 'he asked his wife to give up Coatalen, but she said she would rather leave her own house. He met Coatalen and some violence took place, in fact Van Raalte "thrashed" Coatalen.'⁴ Van Raalte later insisted that he would not agree to divorce Iris until Louis agreed to give her the same amount of money as he had given when he had married her. Although Noel petitioned for divorce in 1918 this seems to have been dropped and he subsequently applied for a divorce through the Scottish Divorce Court, as it was hoped this would be more discreet. Although Louis and Olive were not divorced until 1922, they were clearly separated in 1919 as their house and contents in Wolverhampton were put up for auction in March that year.

The new couple, Louis and Iris, appear to have taken up residence in Paris towards the end of 1919, whilst retaining a flat in central

Spot the family likeness? This photograph was taken in 1917 at the Bath family home, The Folley. On the left is Mrs Olive Bath (Coatalen's mother-in-law) with her son Henry Bath. The children standing in the foreground are Hervé Coatalen and Gonda van Raalte. On the right two nannies hold the babies Jean Coatalen and Charmian van Raalte.

London for their frequent trips to England. They were finally able to get married on 31 January 1923.

As already mentioned, in 1919 Coatalen possessed a large steam yacht *Zaneta*, on which he lived part-time when it was moored in the Seine but he sold it again in 1923. Around the same time he had acquired the old chain-driven Type 18 Bugatti that had belonged to the French fighter pilot Roland Garros who had been killed in 1918. What his interest in this vehicle was remains a mystery, although one might conjecture that it permitted him

---

3. Marjolie Coatalen, p. 18.
4. Undated press cutting, 'Thrashed his Wife's lover. Midland co-respondent in a Scottish divorce suit'.

Chapter Five

The marriage of Louis Coatalen to Iris was featured in *Tatler* in February 1923 without mentioning her previous marriage. The photograph of Iris with a long string of pearls was by Malcolm Arbuthnot.

to test reversed quarter-elliptic springing which became a feature of the little Talbot 8 under development at that time. He sold the Bugatti to Ivy Cummings in 1922 and she named it *Black Bess*. As a workaholic he spent his time shuttling between England and France. He divided his time between the flat he rented at 16 rue Spontini in Paris and another at 93 Lancaster Gate in London. He also apparently kept 'a room' in Wolverhampton for his visits to the Sunbeam factory.

Six

Blowers and the Seeds of Destruction 1924–29

Although it was not evident at the time, 1924 may be seen as the watershed moment when the long-term fate of the STD business was sealed. The group was short of funds again and it issued £500,000 worth of high-interest ten-year guaranteed notes at the beginning of the year. As time passed it was this event, combined with insufficient profits, that came to discolour policies and progress. However, it took a while for this to have an effect, and to start with the challenges of the new developments were all-absorbing and exciting. The racing cars were getting faster and faster, breaking the 150 mph and then the 200 mph barrier.

1924: The Introduction of Supercharging

Early in 1924 Sunbeam was developing a new 3-litre sports car to take part in the Le Mans 24 Hour Endurance race. The rules for this event, being held for the second time, demanded that cars taking part had to be standard models as catalogued and sold to the public, and cars in the 3-litre class would have to cover 1,230 miles, which meant achieving an average speed of 51 mph. This was the opportunity to demonstrate how racing car developments could feed directly into the design of a fast touring car. A six-cylinder engine with twin overhead camshafts inspired by the Grand Prix cars was produced but with a cast iron cylinder block for production in place of the fabricated blocks of the racing cars. In April a prototype was on the road but the initial design of the camshaft drives proved unsatisfactory. The Sunbeam entry did not appear at the race at Le Mans in June, as there was insufficient time to make the necessary changes while at the same time developing cars for the Grand Prix. Le Mans that year was won, for the first of many times, by a Bentley. The subject of the Three-Litre Super Sports Sunbeam then went quiet until the following year.

Chapter Six

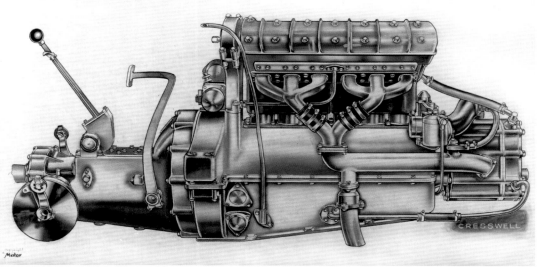

Top: Cresswell drawing of 1924, supercharged 2-litre, six-cylinder Sunbeam GP engine showing inlet manifolding.
Bottom: Original Sunbeam works photo of 1924, 2-litre, six-cylinder Sunbeam GP engine with supercharger mounted at the front and gearbox complete with brake servo at the rear.

It seems likely that development of the cars for the Grand Prix was not straightforward either, as Coatalen and his engineers were moving into an area of new technology – supercharging. In April, around the same time the sports car prototype was revealed, Coatalen was pictured in the Wolverhampton works alongside General Huggins, Chairman of the STD Technical Committee, H.M. Stevens, Assistant Chief Engineer, and Captain Jack Irving, head of the experimental department. They were carrying out tests on a specially constructed device to determine the amount of power absorbed at various speeds and at various pressures by a supercharger.[1] The prototype Sunbeam Grand Prix car was not seen in public until May, when Resta tested it out at a couple of hill climbs and on its first appearance it misfired and popped badly. For the first time in Europe a supercharger was installed between the carburettor and the inlet manifold, and once various teething problems had been overcome it was found to give spectacularly better output than the arrangement where the supercharger had blown into the carburettor. Compared to the unblown engine it was said to give 35 per cent more power across its whole range. The detail of this development, which seems to have been mainly the work of Captain Jack Irving, was kept secret but, although the bonnets were kept shut until the Grand Prix in August, gossip about the new system soon spread around the paddock at Brooklands when one of the cars was tested there. By the time of the race in Lyon, the engines had been mounted in longer, lower chassis with new bodies, new radiators, and also fitted with different gear ratios to suit the course. For the first time the Sunbeams were genuinely viewed as the fastest cars in the field, which included serious opposition from Fiat, Alfa Romeo, Delage and Bugatti.

As usual the Sunbeam team was installed modestly and discreetly in a small hotel in the village of Orliénas not far from the circuit where they could work on the cars out of the public eye. Coatalen was seen looking relaxed and eating calmly with his drivers around a small table on the pavement at the exterior of this establishment before the race. This was in notable contrast to the Fiat team whose

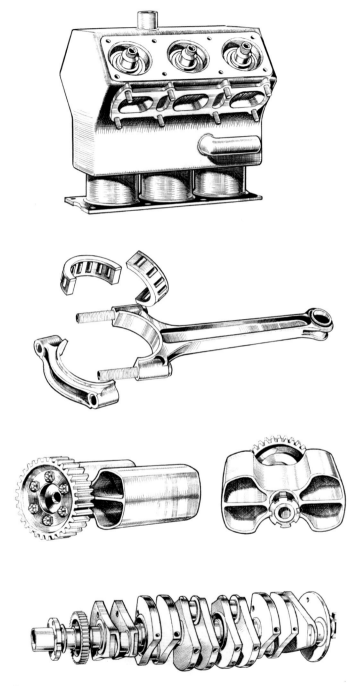

From top:
Cresswell drawing of cylinder block
Connecting rod and split cage roller bearing
Supercharger rotors
Crankshaft

1. *The Autocar*, 25 April 1924, p. 741.

Louis Coatalen stands beside the first of the newly completed 1924 2-litre, six-cylinder, supercharged Grand Prix Sunbeam in Wolverhampton. At the wheel is Dario Resta, with Bill Perkins.

headquarters were in a château, and Bugatti who had installed an enormous marquee at the circuit in which he camped (or glamped?) in style. During the practice period Bugatti and Coatalen engaged in an 'animated conversation' with Bugatti being opposed to supercharging and maintaining he could get the same results with his eight-cylinder atmospheric pressure engines. The 14-mile circuit itself was challenging and potholes had been simply filled with earth so that, when Coatalen accompanied one of his drivers for a lap, he declared on his return that 'every organ in his body seemed to have been shaken out of place'.[2]

In the race the most serious competition came from Alfa Romeo, Fiat and Delage, but Segrave on the Sunbeam led for the first three laps until he dropped back with ignition problems. Guinness was up with the leaders occupying second or third place until a bearing failed in the transmission on his twenty-first lap. Resta, with the third Sunbeam, also had ignition problems which delayed him, and he finished down in ninth place two laps behind the leaders. Segrave had battled on, even putting up the fastest lap in 11 minutes and 19 seconds (76.25 mph) and he finally finished in fifth place. Frustratingly, it emerged that the ignition problems were due to new Bosch magnetos which had been fitted just before the race.

A month later Resta set out to break short-distance records at Brooklands with one of the 2-litre Grand Prix cars. He established a number of International Class E records up to 5 miles (including the flying start 1 mile at 119.56 mph) but a tyre came off and in the accident which ensued Resta was killed and Bill Perkins, who was riding as mechanic, was injured.

Meanwhile, the Talbot-Darracq racers had also been fitted with superchargers and appeared a few weeks later at the 200 Mile Race at Brooklands.[3] George Duller took Resta's place in the team alongside Guinness and Segrave and the little 1,500 cc cars dominated the race again, finishing in the order determined

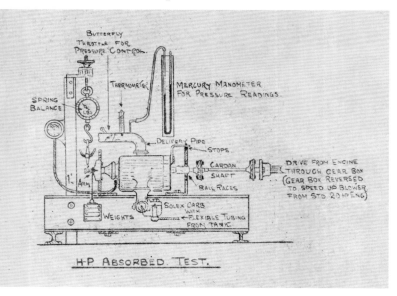

Works diagram of the apparatus used to measure the horse power absorbed by the supercharger drive.

2. *The Autocar*, 25 July 1924.
3. The Talbot-Darracq victory in the Swiss Voiturette Grand Prix held in June 1924 when Kenelm Lee Guinness came first and Dario Resta second seems to have been a final outing for the 1923 un-supercharged cars.

Blowers and the Seeds of Destruction 1924–29

Top left: At the Grand Prix de l'Europe, Lyon, in August 1924. Left to right: Tommy Harrison (foreman racing mechanic) Dario Resta (race driver), Louis Coatalen 'Le Patron' (Team Director) and Vincenzo Bertarione (race car designer).

Top right: Lyon, 1924. Sunbeam mechanics preparing Guinness's car. Note how close the distributor cap of the magneto is to the exhaust pipe. The following year the magnetos were mounted under the scuttle driven off the back of the camshaft. Left to right: Murray, Broome, Atfield, Perkins, MacDonald (?), Harold, Barrett, Moriceau, Taylor, Ridley.

Bottom left: Lyon, 1924. The Sunbeam team before the race. Left to right: Harrison, Grattan (tyres), K. Lee Guinness, Segrave, Murray (behind Segrave), Resta, Perkins, Atfield, Ledu, Bertarione, Marocchi, Miss Irving, Ridley, Irving, Harold, Barrett, Broome, Bill, Taylor, MacDonald. Seated in car: Scales and Moriceau.

Bottom right: During the 1924 Grand Prix at Lyon, Louis Coatalen, standing on the counter and holding the signalling flags, supervises Segrave's pit stop. Segrave bends over to pick up a can of fuel to tip into the funnel on the tail while his mechanic changes a rear wheel.

Chapter Six

Dario Resta and Bill Perkins in the Grand Prix Sunbeam before the accident at Brooklands during a record-breaking attempt on 3 September 1924 in which Resta was killed and Perkins was injured, suffering severe burns.

by lots drawn before the start, Guinness first, Duller second, Segrave third, at a remarkable average speed of 102.27 mph which went some way to restoring the team's morale. In early October Guinness and Segrave were entered for the San Sebastian Grand Prix in Spain with the 2-litre supercharged Sunbeams where once more they were dealt a cruel blow. The Lasarte circuit was heavily cambered and rough – De Alazaga, a private entrant on an old Sunbeam, crashed in practice. Officials gave orders for sand to be spread on the track, but clay was used as a substitute with the result that when it rained it became even more slippery and it was like driving on butter. Guinness skidded and hit a boulder, which projected the car across the road and into a wall. Both Guinness and his mechanic, a young man called Barrett, standing in for Perkins who was still recovering from his injuries sustained at Brooklands, were thrown out of the car and down an embankment. Barrett was killed and Guinness injured. Segrave stopped but was reassured by officials that his team mate had survived, so continued the race and finished as the winner. On the one hand the Sunbeam had fairly beaten the competition of Bugattis and Delages but because of the circuit and the wrong choice of ratios, Segrave reported he had gone throughout the race in second and third gears. Even so the gearbox had withstood six hours of racing and nearly 400 miles of this treatment. The death of Tom Barrett and the injuries sustained by Guinness, which resulted in him giving up racing, cast a long shadow and led to all celebrations being cancelled.

A couple of weeks later, in mid-October the three supercharged Talbot-Darracqs were racing at the new Montlhéry circuit near Paris and Segrave was once more at the wheel. For the final outing of the year the other drivers were Jack Scales, head of the Talbot-Darracq test department, and Bourlier. Each one of them led the race at some point but they finished in the order Scales, Segrave, Bourlier, and the winner's speed was 100.3 mph.

It is interesting to note that Talbot-Darracqs equipped with twin overhead camshaft engines had also been taking part in touring car competitions during 1924. These were the mysterious 68 x 103 mm, 1,496 cc engines fitted into touring chassis which may have been prototypes of an engine intended for production, but which were never offered for sale.[4] Georges Roesch, who was working in Suresnes at the time, witnessed the difficulties that arose in trying to put this engine into production. 'It proved impossible to manufacture the French engine without so much hand-fitting and individual attention that money was lost on all sides; for in contrast to the Sunbeam design (for the twin overhead camshaft Three Litre), little effort had been made to civilise them.'[5] However, they ran successfully in events as diverse as Shelsley Walsh Hill Climb and the Circuit des Routes Pavées near Lille.

The decision by the British Government to drop the 33 per cent McKenna import duty on cars raised hopes that it might be possible to increase sales of Talbot-Darracq cars in Britain but Owen Clegg was more cautious. Due to the scarcity of labour, the cost of production was rising in France and violent fluctuations in the value of the Franc were not helpful to trade. He was quoted as saying, 'in France we have just dropped from a war of prosperity to a period of depression'.[6] Clegg's caution was justified when the duty was re-imposed the following year.

As a distraction from preparing the cars for the year's motor racing programme, Coatalen found himself jousting with W.O. Bentley through the correspondence columns of *The Autocar*. Bentley had launched an attack in March 1924 in which he wrote at length about his view that the two oft repeated statements, 'racing improves the breed' and 'the racing car of today is the touring car of tomorrow', were both 'equally and utterly untrue'. He continued,

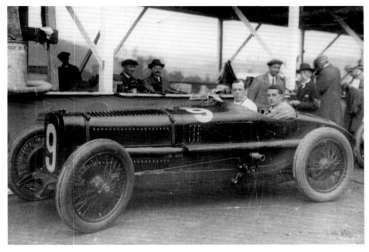

San Sebastian Grand Prix, 1924. Kenelm Lee Guinness with Tom Barrett in the 2-litre, six-cylinder, supercharged Sunbeam before the race in which Barrett was killed and Guinness injured. Guinness never raced again after the accident. Segrave, in a sister car, finished in first place.

'I have always held that the racing of a "standard car" is the only way in which its weaknesses can be exposed so that its design may steadily be improved.' He admitted that the Sunbeam success in the 1923 Grand Prix was a great achievement that benefitted British technical prestige but the only utility of 'an unbeatable racing car is a magnificent, if ephemeral, form of advertisement'. Predictably, Coatalen felt obliged to defend himself. He wrote that he 'very strongly disagreed with the views expressed by Mr W.O. Bentley' and that he was very much surprised 'that he should have gratuitously started this argument'; he tried to put him in his place by noting he was one of the youngest manufacturers and therefore of limited experience. Secondly, Coatalen pointed out that the Bentley 'is practically a standardised form of special racing car as used in the 1914 Tourist

---

4. See Alain Spitz, *Talbot*, p. 192.
5. Anthony Blight, *Georges Roesch & the Invincible Talbot*, p. 21.
6. *The Autocar*, 9 May 1924.

Top: Jack Scales finished first in the inaugural races at the Montlhéry Circuit in October 1924 driving a four-cylinder, 1½-litre, supercharged Talbot-Darracq. His team mates Segrave and Bourlier came second and third.

Bottom: Jack Scales and Jules Moriceau after winning the Circuit des Routes Pavées on a Talbot-Darracq DC Sport in 1924.

Trophy race' and that it incorporated many features that were the result of the short cut in research provided by racing. Bentley's response, while claiming to be 'pleased and flattered to have such a distinguished and well-equipped antagonist' disagreed with this analysis. He insisted that the 'issue is purely academic' and he had no intention to belittle the cars with which Coatalen was connected and even admitted 'we may be both right or both wrong'. But he included a particularly personal barb about Coatalen's aero-engine designs: 'Mr Coatalen frequently quoted the influence of racing car design upon aircraft engines. Previous to the war he had a unique racing car experience. This should have enabled him to produce a predominant aircraft engine. Did it?' This particular subject was subsequently ignored but another reader came to Coatalen's aid by pointing out that the Bentley instruction book included the statement that the Bentley 'contains many features which have hitherto been associated with racing cars in the mind of the average motorist'. Coatalen's final response was that 'the burden of my argument – which was simply a defence against Mr Bentley's attack – was simply that racing car practice accelerated development'. He still held the view that the Bentley was 'very definitely a product of racing experience' even if it was mainly the result of other people's experience. In contrast, he stressed that the 'makers of the Sunbeam are not content to follow; they want to lead.... My object in racing is primarily the very great research value obtained from it. The advertisement obtained is purely incidental.'

One development that Coatalen had pointed out as coming from racing, and which he claimed showed how Sunbeam was ahead of Bentley, was the use of a torque tube in place of a Hotchkiss drive. Some month's later a correspondent queried why, if the torque tube was so superior, were the Talbot-Darracq racing cars still fitted with Hotchkiss drive? Coatalen's answer was simply that they had combined new engines with the old chassis and that the difference was not sufficient to justify a complete rebuild. Another correspondent wrote to recall that 'the following quotation was attributed to Mr Coatalen about 12 years ago; "if you can take an ounce off my pistons you can put a hundredweight on my chassis"'. Thus, another Coatalen aphorism was added to the collection alongside 'racing improves the breed'.

The number of serious motor racing accidents during the year in which Resta, Barrett and Zborowski had been killed, and Perkins, Guinness and others had been injured, naturally gave rise to discussions about speed and safety when considering the regulations for racing in 1925. Proposals were put forward that mechanics should no longer be carried during races. Although Jack Scales was in favour of this idea, both Segrave and Coatalen favoured the maintenance of riding mechanics. Surprisingly, for a man who had spent his career trying to make his cars faster and faster, Coatalen put the cause of recent accidents down to the cars being too fast. He argued that, instead of dropping mechanics, it was much more desirable to reduce speeds by making body sections larger, thereby increasing wind resistance. Having monoplace cars would make things more dangerous by increasing speeds and have very little bearing on touring car development. He even went so far as to state that streamlining had no application in touring car design whereas increasing wind resistance would lead to mechanical progress, as engineers would seek greater horsepower. One wonders if he was dreaming of a return to the early Tourist Trophy races where the cars competed with large 'billboards' fitted to provide resistance? Finally the Automobile Club de France decided not to allow mechanics to be carried in 1925 but bodies had to be a minimum of 31 inches

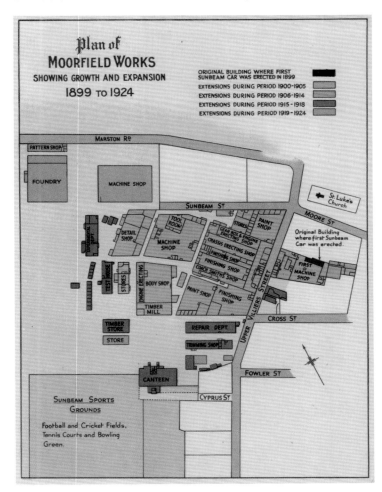

This illustration from the Silver Jubilee booklet shows the huge growth of Sunbeam's Moorfield Works in Wolverhampton before World War I and the position of the foundry and machine shop added post-war.

wide at the level of the seats and engine capacity would continue to be limited to two litres.

Coatalen's distance from day-to-day management of the Sunbeam works is evident from the fact that he wrote from his London address, 69 Piccadilly, in June 1924 to chide works manager C.B. Kay about the lack of progress in replacing the silent chains (used for driving the camshafts on the production cars) with helicoidal timing gears which had been on test for months. He pointed out that Rolls-Royce had adopted them for their 20 hp and said, 'surely we are capable of finding a solution for our motor.' He then went on to say that 'I should warn you that I have heard complaints about the Sunbeam cars, that they are not turned out as they used to be.... I cannot but lay great stress on this point as our good name is the only serious asset we have in the market.'[7] Kay immediately circulated the comments to W. Iliff, Gen. Huggins, Mr Stevens, Capt. Irving and Mr Cozens. It appears, from information Kay obtained from the Fitting Department, that they were experiencing problems with faulty clutches, noisy tappets, steering column rattle and squealing brakes. Coatalen was expected in Wolverhampton the following week when these matters, along with the Three-Litre and the 14 hp low frame, were to be discussed. Kay also provided a summary of orders received and the number of cars delivered in the first three weeks of June. This revealed that they had delivered a total of 116 cars (of which fifty-six were 14 hp and forty-seven were 20 hp) but only fifty-one cars had been ordered.

1925: Sunbeam beats Bentley at Le Mans

The year opened with abortive attempts to set up a 24-hour record at an average speed in excess of 100 mph which Louis Coatalen was of the opinion should be achievable using one of the Grand Prix Sunbeams at Montlhéry. However, sleet and snow in February meant the attempts had to be abandoned. In March things looked up when the Talbot-Darracqs triumphed again at the Grand Prix de Provence finishing first, second and third on the rough track despite the fact they were not fitted with superchargers on that occasion.

7. Letter from Louis Coatalen to C.B. Kay, 19 June 1924 (author's collection).

Top: The two Sunbeams and the two Talbot-Darracqs before the 1925 Le Mans 24-Hour Race.

Bottom: No. 16, the Sunbeam driven by Jean Chassagne/Sammy Davies which finished second at Le Mans soon after completion outside the factory in Wolverhampton. The running boards and wings were changed for the race.

Blowers and the Seeds of Destruction 1924–29

Henry Segrave at the wheel of the Three-Litre Sunbeam which he shared with George Duller in the 1925 Le Mans Race. The early part of the race had to be completed with raised hoods. Segrave led for eleven laps but then retired.

The revised Three-Litre Sunbeam sports car was revealed to the press in March, who opined that 'Mr Louis Coatalen, the designer of the car, has really produced something exceptional in the way of automobiles'.[8] Two of these cars were entered for the Le Mans 24 Hour Race held in June. Henry Segrave, who was sharing his car with George Duller, set off at a storming pace and led the race for 11 laps but subsequently had to retire with a seized clutch. However, the other car, driven by Jean Chassagne and S.C.H. Davis, ran on for the whole 24 hours and finished in second place, having covered 1,342 miles at an average of 55.9 mph. For Coatalen it must have been particularly satisfying that both Bentleys retired when Sunbeams had been able to demonstrate that a touring car, directly descended from motor racing technology, was a viable proposition. In contrast, the two 1,500 cc Talbot-Darracqs retired.

In April, Coatalen wrote to *The Autocar* reiterating the proposal that he had first mooted in 1923, that 'Brooklands could be made both more interesting to the racing spectator and a more valuable testing ground to the automobile engineer' if a chicane was installed on the Finishing Straight. A drawing illustrated how and where this could be achieved simply and inexpensively. 'I need hardly say that the reduction in speed which this would mean ... would be an excellent test for braking, for the gear box, for acceleration, and also personally for the driver',[9] and would also provide an interesting spectacle for the public. By then the Junior Car Club had developed similar plans for their high-speed reliability trial to be held in May.

The French Grand Prix in July 1925 was held at Montlhéry, so for the first time ever it

8. *The Motor*, 31 March 1925.
9. *The Autocar*, 10 April 1925.

The start of the 1925 French Grand Prix held at Montlhéry. The Sunbeams of Segrave (no. 1) and Masetti (no. 7) are followed by Divo's Delage (no. 6).

Two of the three Sunbeams entered for the 1925 French Grand Prix at Montlhéry.

was held entirely on private roads rather than on public roads closed for the event. It was also significant because no mechanics were carried and the distance had been extended to 1,000 km. Three Sunbeams were entered to be driven by Segrave, and two Italian Counts, Masetti and Conelli. The car in which Guinness had crashed the previous year had been rebuilt with a new chassis. Segrave had tried it out

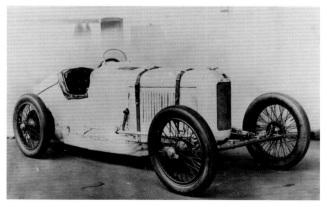

The primrose yellow, single-seater Talbot 1500 cc that set up eight Class F records at Brooklands in August 1925 driven by Henry Segrave.

in hill climb events at Kop Hill and Shelsley Walsh before it went to France where it was driven by Conelli. All three were essentially the same as the previous year but fitted with larger petrol tanks to extend their range. Despite work done over the winter months to improve their performance, they were outpaced by the new eight-cylinder P2 Alfa Romeos which dominated the race from the start. Tragically, Ascari at the wheel of one of these Alfas, crashed and died from his injuries. The other members of the team then withdrew in sympathy, leaving victory to one of the V12 Delages. Two of the Sunbeams retired from the race; Conelli went out as his brake servo failed and Segrave's car suffered a broken inlet valve. However, Masetti worked his way up to second place, which he held until near the end when he was passed by Wagner's Delage, so he finished a respectable third, ahead of no less than five Bugattis. On the strength of that performance

The Talbot-Darracq team drivers for the 1925 200 Mile Race held at Brooklands. Left to right: Guilio Masetti, Henry Segrave and Caberto Conelli. Segrave won at 78.89 mph, closely followed by Masetti. Conelli retired.

Masetti took one of the Grand Prix Sunbeams to Switzerland in August where he broke the record at the Klausen Hill Climb. However, he was less fortunate when he entered it for the Spanish Grand Prix in September, having to retire with a steering problem.

During 1925 Talbot in London was very short of work and it was decided that a record-breaking stunt might generate useful publicity. To this end the Talbot works built a light single-seater racing car which was notable for its tubular front axle and quarter elliptic springs. It was fitted with one of the Talbot-Darracq supercharged 1½-litre engines (although it was suggested it was a Talbot UK engine) and with it Segrave was able to set up eight Class F international records at Brooklands at the end of August. Almost a month later, the 200 Mile Race was held again at Brooklands; the team of three Talbot-Darracqs was entered and once more dominated the race. Segrave led his team mates, Masetti and Conelli, from the start until slowed by a puncture after which Masetti took the lead until Segrave caught up again and they finished first and second less than two seconds apart. Conelli had retired with a back axle problem.

Aside from motor racing, normal production had continued at the three STD factories through 1925. One lesson from racing that Coatalen applied to the production cars was to seek to lower the centre of gravity as much as possible. Wilfred Gordon Aston wrote that the centre of gravity of the 14-40 hp Sunbeam had been dropped by 3 inches without reducing ground clearance 'by virtue of well considered modification of design' (the chassis frame was lowered and the track was widened) and that as a result the centre of gravity was only an inch higher than that of the 1924 GP Sunbeam.[10]

The Talbot factory in London was losing money and it was decided that the new Sunbeam Three Litre Super Sports should be, at least partially, manufactured there and a commitment was made for a large batch to be made even before the car was revealed to the press. The Sunbeam management from Wolverhampton had been drafted in to help with trying to turn around the problems at Talbot. William Iliff took on the role of Managing Director for a while before handing over to Brigadier General Huggins who did the job for eleven

10. Wilfred Gordon Aston, *The Autocar*, 1 May 1925.

Top: Henry Segrave at the Moorfield Works in the V-12, 4-litre Sunbeam soon after it had been completed and not yet painted. In this car on 16 March 1926 he set up a World Record for the flying start kilometre at 152.3 mph at Southport sands.

Bottom: The 4-litre, supercharged V-12 engine of the record-breaking car.

months before also returning full time to Wolverhampton. During the year, Coatalen attended six of the ten meetings of the Talbot directors and when it was announced that the 12-30 Talbot would in future be offered with the option of four-wheel braking, it was said that this system had been designed by Coatalen. In France, Talbot-Darracq introduced a couple of new models and, with an eye to a potential new market, entered a two 12 hp cars in an extremely tough, twenty-day reliability trial across some 5,000 km of Russia; a Sunbeam was also entered. Jules Moriceau, who drove one of the cars, reported back that it had been appallingly severe; the whole of Russia only had about 1,000 km of proper roads and for the rest it had consisted of a sort of cross-country scramble through swamps and sand. Although the Talbot-Darracqs survived well, the market was not attractive financially, as selling cars there would involve giving the Government very long credit.

1926: The Beginning of the End of STD Motor Racing

As 1925 drew to a close, significant changes to the STD motor racing policy emerged as a result of the Group's financial difficulties. The Chairman, James Todd, announced to shareholders at the Annual General Meeting that Sunbeam would not be building any new racing cars in 1926 as 'all the experience required for the next year or two has been obtained'.[11] With the Grand Prix formula limited to cars of just 1,500 cc capacity it was logical to concentrate competition efforts at the Talbot-Darracq works in Suresnes where Bertarione was designing a new car with a straight-eight engine. This sort of limitation was a challenge to Coatalen's ingenuity and already three months earlier it had become known that Sunbeam was building a 'small' record-breaking car – not a racing car. Henry Segrave records that 'Mr Coatalen expressed an interest in seeing what was the maximum speed which could be obtained with a racing car outside the limit of capacity ... of the last few years, but at the same time of quite reasonably rated horsepower ... something comparable

---

11. *The Motor*, 16 February 1926. report of James Todd's statement to STD Motors A.G.M.

Blowers and the Seeds of Destruction 1924–29

to ordinary touring car standards.'[12] This became the 4-litre, 300 hp, supercharged V12, with which Segrave was able to set the World Land Speed record at 152 mph in March on Southport sands. It was a notable achievement for such a light vehicle with a compact engine, as the previous record had been set by Malcolm Campbell at the wheel of the 'old' 18-litre, 350 hp Sunbeam. But it was not achieved without difficulty, as the first attempt had to be abandoned when the supercharger casing distorted, and it cracked again during the run in which the record was set.

Within weeks Parry Thomas, with his 27-litre Liberty aero-engined car *Babs*, had taken the record over 170 mph and thereby had pushed it

Above, left: 1926 eight-cylinder Talbot-Darracq Grand Prix engine under test early in the year. It appears to be unsupercharged (Hervé Coatalen).

Above, right: 1926 Talbot-Darracq body and chassis under construction (Hervé Coatalen).

Right: Vincenzo Bertarione, designer of the 1926 Talbot-Darracq Grand Prix car, standing beside the newly assembled car at the works in Suresnes with Albert Divo at the wheel. Note the ladder-frame chassis for the second car behind.

beyond anything the 4-litre Sunbeam could hope to challenge even when rebuilt with new superchargers. Immediately after Thomas had set up his new record, Coatalen and Segrave had a discussion about the possibility of producing a car capable of breaking the 200 mph barrier. By the end of May 1926 Coatalen had come up with the concept of what was to become known as the 1,000 hp Sunbeam. Because of the lack of

---

12. H.O.D. Segrave, *The Lure of Speed*, Hutchinson, 3rd impression, 1928, p. 183.

Chapter Six

1926, 200 Mile Race at Brooklands. The Talbot-Darracq team consisted of Henry Segrave (in car) Albert Divo (cap with goggles) and Jules Moriceau (right).

a budget to build new engines it had to involve re-using existing power units. There were two World War I 'Matabele' Sunbeam-Coatalen aero-engines, which had previously been converted to direct drive for use in the speed boat *Maple Leaf VII* in 1921, and which were still in storage at the Wolverhampton works. It was decided to mount one at the front and one at the rear of a chassis, both driving into a central gearbox with final drive by chains to the rear wheels (the rear engine had to be modified to run in the opposite direction to the front engine). With a combined capacity of nearly 45 litres producing 871 hp at 2,000 rpm these were considered adequate for the purpose. But Segrave then had to take on the task of raising sponsorship to enable the project to be realised, as Sunbeam would only commit itself to providing the engines and building the chassis.

Meanwhile, work continued on the construction of the team of exciting new straight-eight supercharged Talbot-Darracq Grand Prix cars in France. Although the concept of these long, low cars had been known since the latter part of 1925, full details of the unusual ladder-like chassis side rails and the off-set drive train only emerged in May 1926. Owing to a strike at the Suresnes works they were not ready for the Grand Prix de l'Automobile Club de France at Miramas (which had just three starters) or the European Grand Prix held at San Sebastian in July. They made their debut at the first British Grand Prix held in August on a modified Brooklands track much as Coatalen had proposed three years earlier. The Talbot-Darracqs were faster than the Delages, but painted grass green they were pretending to be English Talbots. They were far from fully prepared so, when Coatalen arrived in person to superintend operations, he sent for more mechanics to help with last-minute adjustments. The drivers were Segrave, Divo and Moriceau. Early on, Segrave put up the fastest lap of the race but suffered from his brakes locking on, misfiring and eventually the car caught fire, so he retired shortly after. Divo's engine also misfired which led to his retirement, but Moriceau suffered the indignity of his front axle collapsing.

By the time of the 200 Mile Race in September the front axles had been replaced, the brakes re-designed and they were altogether more stable. Honour was re-established when Segrave finished first, Divo second. Moriceau cane fourteenth, having lost half an hour by going aground on one of the sandbank corners.

While waiting for the straight-eight Talbot-Darracqs to be completed and to keep the Sunbeam flag flying, Segrave entered the 4-litre, twelve-cylinder Sunbeam for a number of events which slowly revealed its potential as an all-rounder but also highlighted its weaknesses. The Spanish Grand Prix in July was open to cars of all types but after leading for four laps the Sunbeam's front axle broke on the sixth lap when lying second. At the Boulogne speed trials in August, fitted with a redesigned stronger front axle, Segrave frightened himself by breaking the speed record on public roads, exceeding 140 mph on the 6 km-long narrow switchback road. This was followed in September by the Milan Grand Prix at Monza, where Segrave led for 120 miles until he retired when the gearbox cracked. These events were all undertaken by Segrave, supported by a small group of works mechanics and Coatalen was not in evidence. The final outing of the year for the 4-litre car was at Gaillon Hill Climb in Normandy, where Albert Divo set up a new record for the course. For this occasion, instead of grass green, it was painted blue and entered as a French Talbot.

1926, 200 Mile Race, Brooklands. Moriceau skidded and ran aground on a sandbank losing much time, so finished fourteenth.

1926, 200 Mile Race, Brooklands. Albert Divo on the eight-cylinder, supercharged, 1½-litre Talbot-Darracq. Divo finished second behind team mate Segrave, who won at 75.56 mph.

Louis Coatalen standing on the counter of the Talbot-Darracq pits at Brooklands, waving no less than three flags, in 1926.

In 1926 a number of changes were made to the range of cars to be offered to the public in the following year. Coatalen was running a big eight-cylinder, Weymann-bodied Sunbeam saloon on the road, revealing a number of minor problems which resulted in improvements being incorporated in the 1927 models. But the most notable modification to the Sunbeam range was the introduction in the autumn of the 2-litre, six-cylinder, 16 hp Sunbeam that would cost less than the 14 hp, four-cylinder model which it replaced. Another important addition was the 3-litre, six-cylinder, 20hp model. The 16 hp had been conceived for production by Talbot but the STD Control Board had decreed otherwise and transferred production to Wolverhampton. At the Talbot Works, following the relocation of Brigadier General Huggins back to Wolverhampton and the removal of Alan R. Fenn as General Manager, two far-reaching changes came about with the promotion of Andrew Robertson from Works Manager to General Manager and the introduction of the Georges Roesch-designed, six-cylinder, 2-litre, 14-45 Talbot. It featured Roesch's distinctive very light pushrods and rockers that pivoted on an inverted knife-edge. According to Blight, it was Coatalen who, in the autumn of 1925, had sent Roesch back to the Talbot Works in Barlby Road from Suresnes where he had been working for a couple of years. Coatalen had approved the design of the 14-45 in the spring of 1926 to fill the gap left by the transfer of what became the 16 hp Sunbeam to Wolverhampton. This marked the beginning of the recovery of the Talbot business. Coatalen only attended one Talbot board meeting during the year. However, it was his recommendation to the board to purchase new gear-grinding machinery that was accepted and he also supported Robertson's recommendation that £5,000 be expended on machine tools.

Despite this evidence of Robertson and Roesch being allowed to go their own way, the central design office in Suresnes still seems to have been responsible for the model policy of both Sunbeam and Talbot-Darracq, subject to being over-ruled by the STD Control Board. We know little of the workings of this design office housed in the Talbot-Darracq works under Louis Coatalen's direct control. It certainly appears that throughout the 1920s the broad concept and general arrangements were worked out there, but the detailed production drawings would then be delegated to the drawing offices of the individual factories leading to outwardly similar cars being substantially different in detail. John Wyer, a Sunbeam apprentice, described what usually happened:

> The chief designer (Bertarione) produces schematic layouts, that is to say, general arrangement drawings in both sections, longitudinal and transverse, which would establish the main parameters, number of cylinders, bore and stroke, connecting rod length, number of main bearings, method of valve operation and so on. In some cases these schemes are very detailed, even to indicating the specification of materials, the exact size of threads, etc. In others they are just the reverse.... The design schemes are then given to a senior design draughtsman who has the job of producing a set of drawings from which parts can be made and assembled.[13]

It is difficult to perceive that there was anything approaching a rational model plan for the different markets in which the factories operated. Policy seems to have been dictated

Louis Coatalen (centre) with Andrew Robertson (left) and Henry Segrave (right) at the Talbot Works in London. Photographed by Hervé Coatalen when a schoolboy.

---
13. John Wyer letter to Anthony Heal, 25 June 1984 (author's collection).

Chapter Six

by opportunism and immediate commercial imperatives, and this impression is reinforced by the stated Sunbeam policy of introducing modifications and new models as and when they were ready rather than waiting for the Motor Show to make changes for the following year.[14] In addition, we know that models were swapped around between the works and had to change their identities.

In 1922 the 8 hp, 970 cc light car having been developed in France was put into production at the Talbot Works in London. It was sold both as a Talbot and, with cheaper bodywork and a different radiator, as a Talbot-Darracq. Other examples of car designs and identities being exchanged to meet needs of the group at the time are the 1924 16/50 Sunbeam, which became the 18/55 Talbot the year after, although it also emerged as the DA 17/75 Talbot-Darracq. Then there was the 16 hp Talbot that became the 1927 16hp Sunbeam mentioned previously.

Evidence of successful designs that were then further developed on both sides of the Channel can also be found. The roots set down initially for the 12 hp Talbot-Darracq led not only to the 15/40 Darracq DS, but also to the 14 hp Sunbeam. They had very similar looking engines of identical capacity. Likewise, the 20.9 hp (75 x 110 mm) six-cylinder Sunbeam introduced in 1926 was a direct parallel to the TL 20/98 Talbot-Darracq that had been initially introduced as the slightly smaller DUS 17/55 model. To quote Pat Durnford, 'Both engines are outwardly similar and can easily be mistaken … even if none of the parts are interchangeable.'[15] It is surprising that no attempt was made to reduce costs by producing standard parts that could be shared across models.

The patents taken out at this period are perhaps another indication of the extent to which Louis Coatalen had risen above the detail of front-line engineering developments but continued to innovate, coordinate and direct. Of the twenty-nine patents taken out by Sunbeam between 1920 and 1926, twenty-three were jointly with other employees at the Wolverhampton works. One has to assume that they primarily reflect the work of those other individuals such as Stevens (eight), Irving (four) and Cooper (four connected with bodywork). They confirm the success of Coatalen's policy of seeking to employ and encourage the best talent.

1927: 200 mph

By the beginning of 1927 the '1,000 hp' record-breaker was nearly ready. During the summer of 1926 Segrave had successfully negotiated support, either financial or in kind, from six companies – Dunlop, British Petroleum, Castrol, KLG, Andre Hartford and Moseley – so that work on detail design drawings had commenced on 17 September under the supervision of Captain Irving. Construction began on 11 November under the control of Works Manager, C.B. Kay in Wolverhampton. Because of the impossibility of testing the car on a track, a special rig had to be built in the works on which it could be run under load.

---

14. *The Motor*, 24 October 1922.
15. Pat Durnford, 'Whodunnit?', *STD Journal*, no. 127.

A wooden model of the car was tested in wind tunnels by Vickers and the National Physical Laboratory which revealed a serious problem of rear-end lift that led to modifications being incorporated in the body shape. At the end of February everything was ready and Coatalen attended the dinner held in London by the various contributors to the project to wish Segrave success with his attempt and to send him on his way. Coatalen remarked in his speech that you could tell when Segrave was really worried because he bit the knuckles of his left hand. Presumably at this point he was not too worried, as all the calculations had been made and everything was prepared. C.B. Kay was interviewed by the Wolverhampton *Express & Star* and paid tribute to the workers across all departments saying, 'no one ever grumbled if he had to miss a meal or lose half a night's sleep' in order to meet the deadline.

It was widely known that the objective was to exceed 200 mph, which was the reason for going all the way to the long beach at Daytona in Florida. This speed was higher than the other contenders, Malcolm Campbell and Parry Thomas, could hope to attain with their existing

Menu for the 'send-off' dinner for Major Segrave and the 1,000 hp Sunbeam held at the Piccadilly Hotel, London on 28 February 1927.

vehicles and so Malcolm Campbell hurried to make an attempt on the Land Speed Record in February. He achieved a speed of 174 mph to retake the record for himself. Parry Thomas was then prompted to return to Pendine to try to regain the record while the 1000 hp Sunbeam was on its way across the Atlantic.

Wooden model of the 1,000 hp Sunbeam used in air tunnel tests, which resulted in modifications to the final design to reduce lift.

Chapter Six

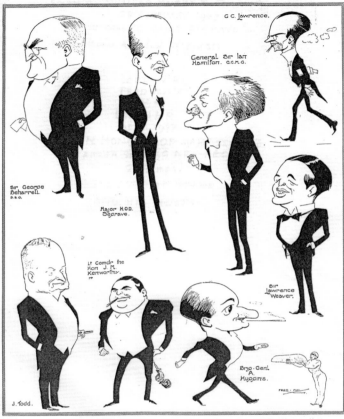

Caricatures of some of the leading figures drawn by Fred May. Segrave is instantly recognisable in the top centre; John Todd, STD Chairman, is bottom left; and Brig. Gen. Huggins of Sunbeam bottom right.

Tragically, one of the driving chains broke during his attempt and Thomas was killed. News of the accident was transmitted to Segrave and the small Sunbeam team on board the liner *Berengaria*, emphasising the risks he was taking in aiming at previously unachievable speeds. Irving would later reveal that the first time the driving chains of the 1,000 hp Sunbeam had been tested under full load they became red hot after less than one minute and as a result, in order to reduce the centrifugal forces at 7,000 ft per minute, special hollow rivets were developed.[16]

Segrave's attempt on 29 March was successful, with a mean speed of 203.793 mph, so that he became the first man to be officially timed on land at over 200 mph. On his return to England, he was greeted as a hero. In his speech at the luncheon given in his honour by the RAC he praised especially Captain Irving who had been responsible for building the car and who was the only one who had been confident from the start that it would achieve 200 mph. It was James Todd, STD Group Chairman, who paid tribute to Louis Coatalen, absent from the celebrations, for conceiving the twenty-four-cylinder twin-engined car. According to the *Birmingham Mail* it had cost £14,000 to build but James Todd was able to reassure his shareholders at the Annual General Meeting that the cost to the Sunbeam Company did not exceed £1,000. So, from the Company's point of view it was an extraordinarily successful and cost-effective exercise.[17]

The next event was the Six Hour Endurance Race held at Brooklands in May for which two Three Litre Sunbeams had been entered by the works to be driven by Segrave and George Duller, with a third one entered by J.W. Jackson. *The Motor* reported that 'it was evident from the very first that the Sunbeam and Bentley teams

---

16. J.S. Irving, 'Building the World Speed Record Sunbeam', *The Motor*, 2 October 1928.
17. *Birmingham Mail*, 30 March 1927, and *The Motor*, 5 May 1927.

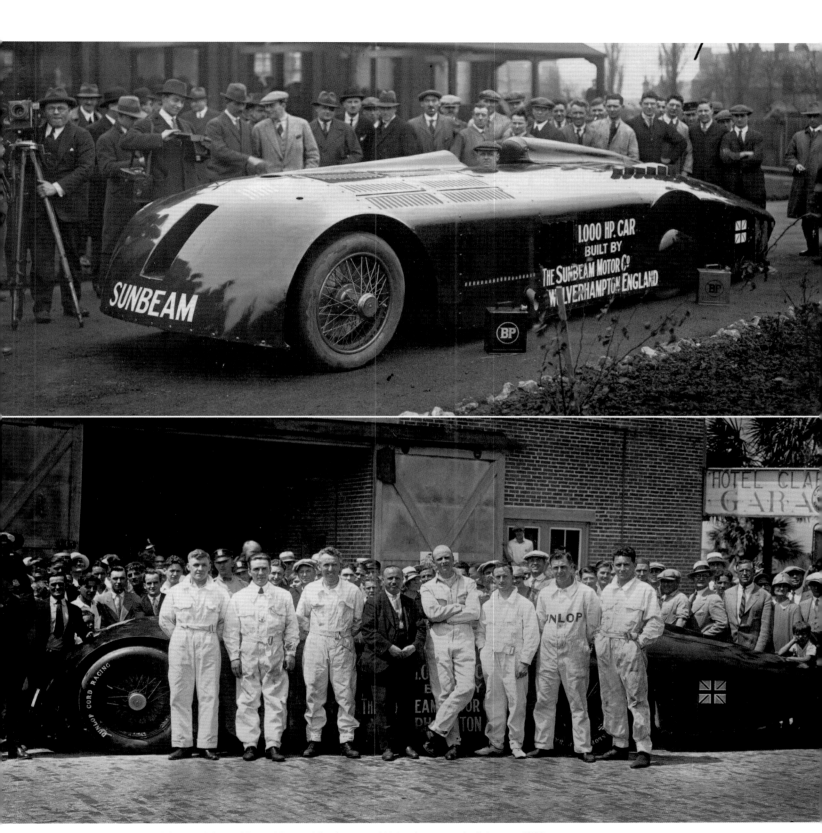

Top: 1927 1,000 hp World Land Speed Record Sunbeam at Wolverhampton in February 1927.
Bottom: The team and car at the Hotel Claremont, Daytona. Left to right: Alec Broome, Bill Perkins, Jack Ridley, Tommy Harrison, Henry Segrave, Edward Lavender, Steve Macdonald (Dunlop) and Dick Slater.

Chapter Six

Henry Segrave in the 1,000 hp Sunbeam on Daytona Beach, with which he set up the World Land Speed Record at 203.79 mph on 29 March 1927.

were really at grips, out to determine once and for all, which was the faster car'.[18] Apart from Birkin's car, the Bentleys had been fitted with experimental duralumin valve-rockers and these failed during the race. George Duller came home first, having covered 386 miles at an average of 64.3 mph, chalking up a convincing win for the Three Litre Sunbeam. Curiously, Segrave seems to have disqualified himself by stopping for petrol somewhere other than the pits and had left the circuit long before the finish of the race. He soon made it known that he was determined to give up motor racing and within three months he had resigned his position as Manager of Sunbeam's London showroom to join the Portland Cement Company as a director and sales executive in connection with concrete roadways. Similarly, Jack Irving left the Sunbeam works to take up another position and thus it became clear that the racing department had been effectively closed, but not before a second supercharged, twelve-cylinder 4-litre racing car had been constructed.

The two 4-litre Sunbeams were entered for a formula libre race held prior to the French Grand Prix and were driven by Williams and Wagner, but both retired with gearbox problems.

In France a certain amount of development of the straight-eight, 1,500 cc Talbot-Darracqs had continued, notably by fitting wider chassis

---

18. *The Motor*, 10 May 1927.

frames. They had practised for the Grand Prix d'Ouverture at Montlhéry in March but five minutes before the start it was announced Divo would not race due to an injured hand. Then at Miramas for the Hartford Cup the Talbot-Darracqs were withdrawn at the last minute, which enraged the spectators who 'rushed onto the track amidst scenes of indescribable disorder' and stormed the Talbot garage.[19]

For the French Grand Prix itself at Montlhéry at the beginning of July, it was the Bugatti team that was withdrawn at the last minute. Although three Talbot-Darracqs started, driven by Divo, Wagner and Williams, both Divo and Wagner retired and only Williams completed the course in fourth place. This limp result marked the end of front-line motor racing for the STD Group. Within a few months the Talbot-Darracq racing department had also been closed down and the cars were sold off. Vincenzo Bertarione, the designer responsible for the racing cars, left at the end of the year to join Hotchkiss, but his friend and colleague, Walter Becchia, remained at Talbot-Darracq.

Towards the end of 1927 *The Motor* published an editorial suggesting that resources should be pooled to produce British Racing Cars in order to avoid repeating the 'pitiable fiasco' of the British Grand Prix when no British car survived for more than a few laps.[20] Over the following weeks the idea was generally welcomed but the many difficulties of realising such a scheme were pointed out by correspondents such as Campbell, Segrave and Guinness. Others were less enthusiastic, such as Herbert Austin, whose view was that 'motor racing was much overvalued' and T.G. John of the Alvis Co. wrote, 'I cannot see any hope of success in your scheme ... when nature bestows gifts on an individual they are almost always accompanied by characteristics which make efficient team work impossible. Thank heaven it is so because it is the best argument we have against Socialism and Communism.' Coatalen, deprived of his racing teams, wrote that he was 'heartily in accord' with the suggestion to pool resources to produce a team of racing cars to hold their own against the world. He pointed out the successes for which he had been responsible and that it was 'not altogether reasonable for the Sunbeam Co. always to shoulder, unaided, the financial load of producing cars to defend British prestige'. He would be glad to help, 'as I consider the value of a series of Continental or International victories is of great importance to British trade'. He suggested that if pooled resources were not achievable, funds should be guaranteed to cars that got through eliminating tests or finished races above certain positions or speeds. Nothing was to come of such ideas until BRM was set up after World War II as a national project and it was many years before that became successful.

Curiously, highly confidential discussions were being held at this period about a possible amalgamation of Sunbeam with the Daimler company. David Burgess-Wise's research in the Coventry Daimler archives unearthed

---

19. *The Autocar*, 1 April 1927.
20. *The Motor*, 18 October 1927.

some correspondence between Coatalen and Percy Martin, from which it emerges that Martin was proposing that they might benefit from linking the two firms. It was Coatalen who put Martin's idea to the STD directors at a board meeting in London in September before returning to Paris for the motor show. The directors asked for time to consider the proposal and they seem to have been in no hurry as it was not until 27 January 1928 that a lunch meeting between Coatalen, his Chairman James Todd, and Sir Edward Manville, Chairman of the BSA group of which Daimler was part, took place. Percy Martin was unable to attend the lunch but followed up the next day, writing to Coatalen, 'I think now it is highly essential that I should have a further talk with you' to propose 'a definite first step to be taken. I think when we do meet it will be a case of you and I agreeing to put a confidential man from each Company together to see what effect our contemplated move would probably have on the business of the several Companies. I am assuming that you feel that most can be gained by agreement on the subject of programme, prices and if possible by merging to a certain extent the sales schemes.' In due course a meeting was set up between William Iliff, Sunbeam's Joint Managing Director, and Algernon Berriman of Daimler in order to discuss details of 'possible arrangements whereby the sales policy of the Daimler Company and the Sunbeam Company can co-operate without coming into conflict' but, as David Burgess-Wise records, 'there the paper trail – and presumably the discussions – ended'.[21] The only hint that these discussions had taken place emerged a year later when James Todd remarked to his shareholders at the AGM that they had received an offer from an unnamed source to buy up an unnamed subsidiary company. As the offer had valued the £1 shares at 35 shillings, Todd was able to imply that the balance sheet was a lot stronger than it appeared and they had no need to consider breaking up the group. However, it seems probable that if Coatalen and Martin had dreamt of rescuing Sunbeam by extracting it from the STD Group, they had come up against the reality of the complications of the Group's interlocked finances. To withdraw Sunbeam would have brought the whole pack of cards tumbling down.

1928: The Beginning of the End

At this point it becomes difficult to discern Coatalen's hand in the affairs of the Sunbeam Motor Car Company. We know that soon after the last known meeting between the Chairman of BSA, Todd and Coatalen, Louis was cruising in the Mediterranean in his new yacht *M.Y. Karen*. C.B. Kay, the man who had worked with him since 1912 and who since 1914 had been Sunbeam's General Works Manager, was appointed to the Board of Directors of the Sunbeam Company and it is tempting to assume this was to fill a gap caused by Coatalen's increasing absences.

The 2-litre Grand Prix Sunbeams had been sold off but continued to appear regularly at

---

21. David Burgess-Wise, *The Automobile*, May 2015.

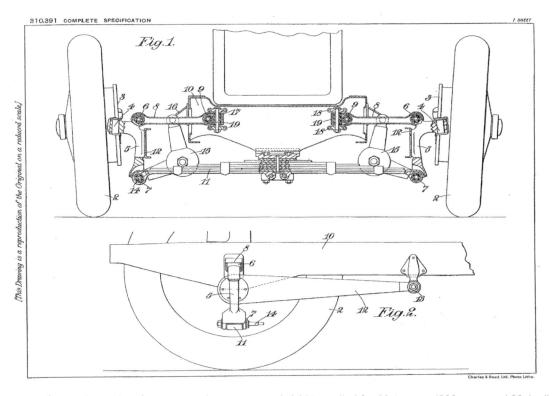

Drawing from independent front suspension patent no. 310,391, applied for 23 January 1928, accepted 23 April 1929.

Brooklands and elsewhere in private hands but with a certain amount of mechanical support from the works. More drastically, the Company decided to diversify into bus production, which seems unlikely to have appealed to Coatalen's instincts. The costs of developing a commercial vehicle range from scratch, combined with plummeting sales of luxury cars pushed Sunbeam into making a loss.

One significant development, however, at the beginning of 1928 was the filing of a patent by Coatalen and Sunbeam relating to independent springing which combined a transverse spring with a pivoted arm from chassis to wheel support, located by a substantial radius rod. In 1932 Walter Becchia was to adapt the design for Talbot in France and it was then adopted by both Delahaye and Delage 'to play a powerful part in the regeneration of the French sports car'.[22] Sunbeam incorporated this independent front suspension in the *Dawn*, which was introduced in 1934. When Anthony Blight, that great proponent of Roesch Talbots, discovered the origins of this design some fifty years later, he was moved to write, 'Interesting that perhaps STD did have some slight importance, after all.'

Shortly before Christmas a new contract was drawn up, signed by James Todd and C.N. Wright, that confirmed Coatalen as Technical

---

22. Anthony Blight, *The French Sports Car Revolution*, Foulis, 1990, pp. 49-50.

Director of Sunbeam, Clément Talbot and Talbot SA, as well as Chief Engineer and Joint Managing Director of Sunbeam for a period of five years. It recognised that he was resident in France and stipulated that he would only be paid by Sunbeam. Was this an attempt to get him to re-engage with sorting out the firm's problems?

1929: A Final Fling

Perhaps it was the result of the new contract but in 1929 Coatalen's influence on affairs was more in evidence again, even though somewhat erratic. At the STD Annual General Meeting in February it was stated that both Arthur Huntley Walker, Managing Director of the group, and William Iliff, Joint Managing Director of Sunbeam, were unable to be present as they were dangerously ill, so possibly there was even less restraint on Coatalen's ideas than usual. Or perhaps James Todd felt the necessity for a smokescreen of activity to draw attention away from the increasingly dire group finances following the negotiations to extend the redemption time on the 8 per cent Notes. Whatever the truth, things began to happen. In May, for example, it was announced that Clément Talbot Ltd. was to build six-cylinder air-cooled engines under S.A.R.A. patents for the new Scotsman car – it seems unlikely that any were actually built for this ephemeral project. Rumours about a new Land Speed Record car started to circulate and were confirmed in July (see the next chapter for the Silver Bullet story). The Three Litre sports tourer Sunbeam was offered with a Cozette supercharger to enable one to be entered for the Tourist Trophy Race.

Less characteristically, the development of buses continued with the introduction of a six-wheeler, double-decker *Sikh* and a four-wheeler, single decker *Pathan* bus. Both were powered by six-cylinder petrol engines (8-litre and 6½-litre respectively) designed by Hugh Rose following his experience at Riley and featured hemispherical combustion chambers with overhead valves at 90 degrees operated by cross-over pushrods from a single camshaft set low down in the block.

Behind the scenes diesel engine development research was taking place, although Coatalen's dream was that these combustion ignition engines should power aircraft rather than lorries, as will be shown in the next chapter.

Private Life

Although Iris returned to London in 1924 to give birth to their daughter Marjolie, she went straight back to Paris and the new baby was only brought out by her nanny, Mrs Holmes, three months later. Iris continued to practise her skill with a rifle and became Champion of France in pigeon-shooting three years running. She was featured in French *Vogue* in November 1926, which published her picture wearing a beige and brown skirt with a man's jacket (English fabric), a large felt hat and thick-soled shoes in yellow leather. She was said to be *'un fusil très réputé'* (a renowned shot) . She took part in competitions in the Bois de Boulogne, Vichy and Aix-les-Bains. The *Cyrano* weekly magazine reported in August 1928 on the event held at Touquet Plage:

il y a des dames qui assistent, très amusées au massacre des pigeons, et même qui y prennent part....Mme Coatalen seule femme ayant gagné un prix de cinquante mille francs, tirant d'ailleurs comme un homme (encore un argument pour les feministes) et qui cette saison, remporte le grand prix des dames.... Mme Coatalen a tué l'oiseau avant même qu'on l'ait vu. Elle revient au stand en caressant lentement le canon de son fusil, comme pour le remercier.[1]

As an illustration of the sort of social circles they moved in, it is notable that among their friends was the very eccentric Princess Violette Murat. Another mutual friend was the artist Augustus John, who described the Princess as utterly fearless and who took him 'to certain places where my sangfroid was put to the severest test'. It was Princess Violette who introduced Augustus John to the habit of taking a teaspoonful of hashish in the form of compote or jam. He recalled falling ill in Ste Maxime,

During school holidays the boys, Hervé and Jean, got to spend time with their father. Here Louis appears to be trying to give his sons a photography lesson in the Bois de Boulogne in front of a sceptical-looking audience.

[1]. Translation: There are ladies who are amused to watch pigeons being massacred and some even who take part.... Mrs Coatalen, the only woman who has won a 50,000 franc prize, firing like a man (another argument for the feminists), carried off the Grand Prix des Dames this season.... Mrs Coatalen killed the bird even before it came into view. She returned to the stand slowly caressing her weapon as if to thank it.

Chapter Six

where Violette and Iris looked after him 'like ministering angels' although he was somewhat unsettled because he had recently witnessed Iris shooting pigeons 'usually, but not always, with deadly accuracy, at Monte Carlo!'[2] In those circles it was fashionable to take opium and both Louis and Iris became addicted. They even had a special room set up in their apartment in Rue Spontini, Paris 16eme, full of Chinese furnishings and lined with black velvet.

During the 1920s Louis' two sons were at boarding school in England. Their mother remarried Colin Hunter Blair but it was their grandparents' home near Wolverhampton that gave them stability. However, holidays were also spent with Louis in France, as is shown by their attendance at the 1923 Grand Prix and Hervé's photos taken inside the Suresnes racing department a few years later. Hervé recollected that during these school holidays with his father they would sometimes have to stop for him to have a smoke of opium.

By the beginning of 1928 Coatalen owned another motor yacht, *M.Y. Karen*, which they sailed into Monaco and Antibes in February that year. By May it was recorded in the Channel crossing between Southampton and Rouen and back.

The marriage between Louis and Iris lasted until 1929 when Iris fell madly in love with Yvonne Franck, a ballerina at the Paris Opera.[3] Iris went to live with Yvonne in a house the latter had been provided with in Montmartre, but although she moved out with her daughter Marjolie, they do not seem to have lived together. Marjolie and her nanny were set up in a house in the country at Jouy-en-Josas that belonged to Princess Murat and so Marjolie only saw her mother very occasionally.

2. Augustus John, *Chiaroscuro*, Jonathan Cape, 1952, pp. 197-200, 178, 243.
3. L'Eventail de Jeanne was a children's ballet choreographed in 1927 by Yvonne Franck and Alice Bourgat for the ballet school of Jeanne Dubost, who had invited ten composer friends (including Ravel and Poulenc) to write a little dance for her pupils. It was taken into the repertory of the Paris Opera and premiered publicly in 1929 (*The Concise Oxford Dictionary of Ballet*, Oxford University Press, 1987).

M.Y. Karen, Coatalen's second motor yacht in 1928.

Seven

Collapse and Recovery 1930 – 39

1930: The Collapse

Coatalen's life hit rock bottom in 1930. His third wife had left him, he made some very poor investments, his World Land Speed Record car *Silver Bullet* was a failure, he was frustrated by the fruitless dealings with the Indian Motorcycle Company in connection with his compression ignition engine patents, and the accountants Price Waterhouse, whom STD Motors had called in to investigate the management of the business, were critical of his role and that of his co-directors. It is perhaps not surprising that Coatalen suffered a nervous breakdown, no doubt exacerbated by his addiction to opium and alcohol, which resulted in him retreating to the island of Capri for about a year. These events are examined in more detail later in the chapter.

Silver Bullet

In early 1929, when both Henry Segrave and Malcolm Campbell were preparing for their World Land Speed Record attempts in America and South Africa respectively, rumours had started to circulate that an un-named, but well-known, British motor manufacturer was considering building a new record breaker for Kaye Don, who had then emerged as the latest successful racing driver. At that stage it was secretive, speculative and no decision had been made. It was not until the middle of the year, after Segrave had successfully retaken the record at 231 mph and duly been given a knighthood, that it was confirmed that Louis Coatalen had prepared preliminary drawings for a car to be built by Sunbeam, which was intended to reach 280 mph. It was confirmed that the drawings had been shown to Kaye Don, who had undertaken to drive it and also that 'an enormous sum of money will be spent on

One of the V-12 engines for the *Silver Bullet*.

the car'.[1] Hugh Rose, who had re-joined the Sunbeam design office in Wolverhampton, was summoned to see Coatalen in Paris in June to be briefed and to receive instructions about working up the detail design. Henry Wilding, in charge of the Experimental Department, who had been Captain Irving's assistant on the previous World Land Speed Record car, was to be responsible for testing and preparing it. By the end of 1929 the two new, 24-litre, narrow V12 engines had been built and were each producing 490 hp, approximately the same as had been obtained from the old Matabele engines. However, it was expected that, once supercharged, the two engines would generate 1,200 hp together. Coatalen visited the works regularly during their construction and was present during the bench tests of the individual engines and again when both had been installed in the chassis and were run together. Henry Wilding recorded that 'we dare not do anything except just what Coatalen instructed'. Although a massive casting encasing four Roots type blowers was initially tried, it was decided to substitute a single centrifugal supercharger that would turn at 17,000 rpm and be driven through a Matteuci gear, all designed by the Italian engineer, P.F. Martinuzzi, Coatalen's new assistant. It proved impossible to make the Matteuci gear work satisfactorily and with only a couple of weeks to go before the car was due to be shipped out to the USA it was replaced with an epicyclic gear drive.

There was just time for *Silver Bullet* to be presented to the Press before being crated up for shipping. Movietone News made a somewhat surreal film of Coatalen appearing through the freezing February fog outside the Wolverhampton works to greet Kaye Don who then departed at the wheel of the Land Speed Record car, towed behind the 4-litre Sunbeam *Tiger*, driven by Bill Perkins. Apart

1. *The Motor*, 2 July 1929.

 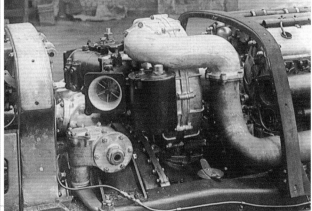

Left: Supercharger casing for quadruple Roots blower.
Right: Centrifugal supercharger and 'Amal' carburettors.

Collapse and Recovery 1930–39

Silver Bullet chassis with engines mounted in tandem fed by single centrifugal supercharger driven from rear engine but supplying mixture to both engines via long pipes.

from revealing that Coatalen's accent was still very thick and French, that Kaye Don was a very stilted performer in front of a movie camera, and that the *Tiger* was an extraordinarily versatile car (it had broken the Land Speed Record and set up the Brooklands lap record but could also be used as a tow car), the film underlines the different expectations that were being put about at the time. Coatalen talked about the engines producing 4,000 horsepower – whereas we know they produced about 900 hp; 300 mph was suggested as a possible maximum speed – while Kaye Don more realistically hoped to achieve between 240 and 250 mph. Sunbeam's own publicity material suggested that *Silver Bullet* was geared to reach 248 mph. Preparation time had run out and this underdeveloped and insufficiently tested special of which too much was expected had to be loaded in a crate and despatched to Daytona Beach in America.[2]

Silver Bullet was shipped from Southampton on 26 February 1930, aboard the luxury liner *Berengaria*, accompanied by the whole support team, which included Coatalen and Martinuzzi. Coatalen went first to New York to try to negotiate a deal on his Diesel aero-engine patents and only arrived at Daytona on 15 March after the first trial runs with *Silver Bullet* had been made. Perhaps the problems he encountered in New York had pushed him over the edge because Henry Wilding recorded that 'my troubles began from the moment he put his foot in the garage.... Mr Coatalen was ill and in

---

2. Moving Image Research Collections, University of South Carolina, Silver Bullet Outtakes, Fox Movietone News Story 5-441, 21 February 1930. www.mirc.sc.edu

Kaye Don posing with *Silver Bullet* at Daytona Beach before an attempt on the World Land Speed Record.

Chapter Seven

A rather depressed looking Kaye Don (left) at Daytona with Louis Coatalen (right) who is discussing the car's problems with his assistant, Martinuzzi (behind Don). (NMM)

my opinion was not fit to be there at all.' Various technical difficulties, including distortion of the supercharger casing, were encountered but one problem about which nothing could be done was the state of the beach itself which was the most serious problem the team faced and prevented the car ever being run near its maximum.

Coatalen left Daytona on 2 April and went back to New York, leaving instructions not to come home until the record had been broken. After a month in Daytona and numerous fruitless runs on the beach, Don and Wilding travelled to New York to get his permission to abandon the attempts. Within three days they were all aboard the *Aquitania*, setting sail for Southampton on 16 April while Tommy Harrison was left to pack up the car and all the spares. He, the car and the other mechanics followed on the *Berengaria* a week later. *Silver Bullet* had never come anywhere near the Land Speed Record speed.

It had cost the company £15,557 to build the car, to which had to be added the cost of the record attempts which amounted to £4,102.[3] Although the build cost was remarkably similar to that reported for Sunbeam's 1,000 hp car three years earlier, for which engines were already in existence, Kaye Don had not raised the same level of sponsorship as Henry Segrave had succeeded in doing on that occasion. James Todd, the Chairman, stated that Sunbeam had received a £10,000 cash contribution and that the attempt was 'made possible by a great friend of K. Don's who does not wish his name to be mentioned at the moment'.[4] Had the attempt on the record been successful, no doubt a gloss would have been put on these figures as representing good value for money. As it was a failure, recriminations started to circulate, with some suggesting that Kaye Don's driving was to blame while from his side it was alleged that the car did not have the performance that had been promised. As if to demonstrate that his courage should not be questioned, soon after returning home Don set up a new Brooklands lap record with one of the 4-litre V12 Sunbeams. Evidence was prepared for the case, Kaye Don v Sunbeam Motor Car Co. Ltd, to be heard at the King's Bench Division in the High Court of Justice but the parties settled before it came to Court

---

3. Sunbeam Works statement of cost sheet dated 30 June 1931 (author's collection). As a further comparison, Captain Irving recorded that Segrave's *Golden Arrow* had cost £11,559 to build.
4. *The Motor*, 1 April 1930.

so a public spat was avoided.[5] According to Henry Wilding's prepared statement of evidence, two members of the design team, Hugh Rose and Mr Harrop, a mathematician, were discharged by the Company in June 1930, thus suggesting that it was felt that the design had been unsatisfactory.[6] What happened to P.F. Martinuzzi, who was responsible for the design of the supercharger and its drive which gave so much trouble and produced so little benefit, is not known. Coatalen was always a great publicist and not afraid of embroidering the truth to make a better story but it appears that with *Silver Bullet* he was seriously losing his grip on reality.

Back in France Coatalen was soon on to another improbable project. It is not clear if it was as a spin-off from his trip to the United States but it seems likely that he had seen the new V-16 Cadillac during the visit. Some time after his return he ordered a slightly used Cadillac and possibly also a brand-new front-wheel-drive car, the Ruxton, which was a short-lived design that had been developed as a prototype with a view to interesting a manufacturer in producing it commercially. Coatalen's idea seems to have been to try to combine the two for production in France. Owen Clegg reported back:

> I tried the Cadillac on Sunday for 100 km. It is certainly a very fine car, but I was very disappointed with the speed, and I do not think the pick-up is what one might expect for a motor of that size. Very dangerous on corners, but, as you say, that could be remedied.
>
> I think we should certainly set about making a front wheel drive experimental car, but 8 cylinders in my opinion would be quite enough without going to 16.[7]

Having examined it closely, the Cadillac was sold on within a couple of months via Gustav Baehr to Mr Michel, the General Motors representative in France and no more was heard of the project.

STD Motors Trading Difficulties

At the Annual General Meeting of STD Motors in early August 1930 it was announced that Mr Louis Coatalen was unable to attend due to 'serious illness'. In order to try to sort out some of the problems at the Sunbeam Works, Mr Andrew Robertson, who had been largely responsible for the turnaround at the Talbot works a few years before, had been appointed General Manager of Sunbeam as part of a major reorganisation. Earlier in the year STD Motors had carried out a drastic restructuring of its balance sheet but with

---

5. Barrie Price recalls that Don told him he received £5,000 in damages from Sunbeam. Perhaps this was a part repayment of the sponsorship money he had raised. Letter from Barrie Price to Peter Morrey, 4 November 2017. My thanks to Peter Morrey for this information.
6. Henry Wilding joined Sunbeam Motor Car Co. in 1910 and since World War I had worked in the racing and experimental department which he had headed up since the departure of Captain Irving in 1927. His various reports on *Silver Bullet* were reproduced in the *STD Journal*, June 1986 and January 1987, and in Anthony Heal's book *Sunbeam Racing Cars*.
7. Letter from Owen Clegg to Louis Coatalen, 20 June 1930.

Chapter Seven

the deepening Great Depression this was evidently inadequate. It was also announced that, at the demand of certain shareholders, the accountants Price Waterhouse had been asked to make a thorough investigation of the management and control of the subsidiary companies in the group. In addition to undertaking an independent management review, they would advise on the possibilities for disposing of one or more subsidiary companies or, alternatively, merging the whole group with another motor manufacturing concern. After Price Waterhouse had submitted its report towards the end of the year, a new Board of Directors was appointed. In his absence this brought to an end Coatalen's twenty-two-year career with Sunbeam, at the same time as ending the STD Chairmanship of James Todd JP, FCA.

Although STD Motors struggled on for a few more years until it was finally taken over by Rootes Securities Ltd in 1934, the cause of its plight at this stage has sometimes been blamed on the expense of Louis Coatalen's motor racing programme in the early 1920s. This is certainly an oversimplification of events. The company did not pay a dividend on its Ordinary Shares from 1920 onwards, so it is clear that from the very creation of the Group, it was not able to generate sufficient profit to satisfy the shareholders or re-invest in the business and its products. To put this into context it is necessary to go right back to the formation of STD Motors Ltd.

The Talbot historian, Stephen Lally, suggests that the group of companies which eventually became STD Motors was initially put together at the end of World War I in order to capitalise on the purchase of war surplus vehicles that could be repaired and sold on.[8] The share capital of A. Darracq Co. Ltd had been enlarged in 1918 to fund the purchase of Heenan Froude, the general engineering company famous for making water brake dynamometers and constructing Blackpool Tower. In September 1919, Darracq, having made another big issue of shares, acquired the entire capital of Clément Talbot Ltd, thus obtaining a London works well equipped and experienced in repairs to ex-military cars and lorries, and the following year the vehicle spring manufacturing business Jonas Woodhead and Sons of Leeds was purchased. On 1 October 1919 a contract was entered into between the Liquidation Commission of the United States War Department and the purchasers who were listed as: Owen Clegg, James Todd, Lt Col George W. Parkinson, Arthur Huntley Walker, Soc. Anonyme Darracq, A. Darracq & Co. (1905) Ltd, Clément Talbot Ltd. Under the contract, the US sold them a large quantity of motor vehicles and equipment, held at various locations in Germany, for £3,250,000. Payment was to be made in eight instalments over six months. By December 10 per cent of the money had been paid but the purchasers then asked for a readjustment of the dates of payment. It was agreed that the outstanding £2,950,000 should be paid in three equal instalments in July and December 1920 and April 1921. Significantly, the group of purchasers was joined at this point by

---

8. Stephen Lally, '1920–1925 Talbot's Years of Neglect', *STD Journal*, Spring 2014.

Sir Percival Perry of The Motor Organizations Ltd, who took charge of the conditioning and marketing of the vehicles. However, James Todd was also a director of that company so he was gaining further experience in large-scale financial deals.⁹

The Sunbeam Motor Car Company, whose founder John Marston had died in 1918, was not necessary to this plan to restore and sell war surplus vehicles, but when it became available a deal was done to amalgamate the two enterprises A. Darracq and Sunbeam, and double the size of the group, so that in August 1920 it had an authorised capital of £3.3 million. According to *Motoring Entente*, the Sunbeam directors 'had not met financiers before and in Todd they decided they had found the figurehead they needed. He was a Lancashire man with a broad accent which increased their confidence in him, and he had a brilliant way of handling meetings.'¹⁰ Shares were exchanged on a one-for-one basis, but one cannot help suspecting that the Darracq valuation was much more speculative than the more solidly founded Sunbeam balance sheet. How much debt did Darracq and Talbot still owe to the US Government and on what basis had their French assets been valued? The French franc had lost 70 per cent of its pre-war value by April 1920 but optimists still believed it would recover when Germany was forced to pay reparations.

Thus from the very start the STD finances were precarious. As already mentioned, the holders of ordinary shares did not receive a dividend payment from 1920 onwards and in 1922 the Preferred Ordinary shares also fell into arrears.

> Some of the trouble which was not revealed at this time, however, concerned the non-payment of considerable sums still owing to Sunbeam's for aircraft work carried out before the amalgamation. Talbot's had meanwhile purchased quantities of war-surplus vehicles from dumps in Flanders for reconditioning ... and as these had never been paid for, as soon as the Group was formed, the Government was able to enter a simple contra in its books, and the Sunbeam assets suffered accordingly.¹¹

By the beginning of 1924, the company was in need of a substantial injection of working capital and had to resort to issuing £500,000 of 8 per cent Guaranteed Notes. Various sources suggest that producing a racing team in the 1920s would have cost between £20,000 and £50,000 a year, so it is clear that racing alone was not what this sum was required for. The Notes were issued by the STD Motors holding company but secured by a specific charge on the whole share capital of

---

9. Sir Percival Perry (1878–1956) had set up Ford in the UK before World War I but became Deputy Controller of the Mechanical Warfare Department of the Ministry of Munitions during the war. For this he was knighted. In 1920 he led the consortium which purchased the Slough military transport depot that then became the Slough Trading Estate. In 1928 he was appointed Chairman of Ford Motor Co. Ltd, and in 1938 was created 1ˢᵗ Baron Perry.
10. Nickolls and Karslake, *Motoring Entente*, Cassel, p. 75.
11. *Motoring Entente*, p. 189. Talbot were acquiring and repairing British ex-military vehicles before the deal with Darracq and the USA was done.

Chapter Seven

the subsidiary companies. In turn, the subsidiary companies created Debentures charging their undertakings and assets as security for these funds. The effect was that:

> The guaranteeing companies found themselves hampered always by their debentures which prevented them from raising temporary capital when it was required. The result of this was a continual shortage of working capital, as those members of the group which at any time had any surplus funds were forced to pass them over to others and consequently were unable to accumulate any reserves.[12]

The high rate of interest on the notes meant 'large sums were required each year to meet the redemption payments which further drained the cash resources of the group'.[12]

As the Group financially stumbled through the 1920s with the share price sinking as low as 3 shillings on occasions, STD Motors remained just profitable but there was never sufficient surplus to pay a dividend on the ordinary shares and only a partial payment was made in 1927 to preference shareholders. It was a difficult period for trade and other manufacturers experienced similar problems. Wolseley, having optimistically raised capital by a similar scheme and then made losses, was taken over by William Morris. Austin also accumulated seven years of arrears on its preference dividends and had to undergo a second major financial restructuring in order to be able to carry on.

Throughout most of the 1920s the Sunbeam factory in Wolverhampton was the most profitable of the STD subsidiary companies. For example, it is recorded that out of the Group total net profit of £135,000 in 1923, £101,000 came from Sunbeam when the racing team was still in full activity. In contrast, the French company, Automobiles Talbot S.A., regularly made losses which by 1930 amounted to £1,457,783, most of which had been made good by loans from other parts of the group. With the French franc further devalued by 50 per cent, the group was eventually forced to write-off about 80 per cent of its £2 million investment in France.

Ironically, by 1930 Sunbeam was losing money and both Clément Talbot in the UK and Automobiles Talbot, France were contributing small profits. James Todd came under mounting pressure from angry shareholders and frustrated trade creditors. Although he had instigated the previous year a scheme to delay payments on the 8 per cent Notes (which were due for repayment in increasing instalments by 1934) over an additional fifteen years, he had to agree not only to the capital reorganisation scheme but also to instruct Price Waterhouse to make the thorough investigation of the management and control of the subsidiary companies referred to already.

In preparation for the investigation by Price Waterhouse, Owen Clegg wrote to Louis Coatalen reminding him that he owed Automobiles Talbot some money and recommending that he should clear up the

---

12. H.K. Newcombe (Chairman), report to shareholders at 29th AGM of STD Motors Ltd, 28 February 1935. I am grateful to Pat Durnford for providing copies of a number of period reports on which this section is based.

balance before the accountants arrived. The schedule he enclosed showed that, over a couple of years, he had accumulated a debt of some £600. Although not a huge sum, it provides an interesting insight into Coatalen's expenditure at this time, as he had used the company to advance payment for deliveries of coal and the rental of his Paris apartment in rue Spontini (Fr 16,000 a year) as well as to pay for train fares and the cost of telephone calls. In the second half of 1929 there were five trips by Pullman to London and a couple of trips to Naples. Clegg wrote, 'I do not want you to think I wish in any way to push you in this matter, and I know that you at one time, along with me, lent considerable sums to the Company when they were hard up, but all the same I think it advisable that everything as far as possible should be "régulier"....'[13]

The Price Waterhouse accountants arrived in Paris at the beginning of August to go through the books of Automobiles Talbot, but they did not finally submit their report on the whole group until the end of 1930. As a direct result of their recommendation the entire Board of Directors resigned in March 1931 and a new Board was appointed.

Personal Investments

A few scraps of information survive that enable one to have some sort of idea of Coatalen's personal finances around this period. It may be reading too much into those papers, but the impression remains of a man who was perhaps too easily talked into taking up rather doubtful investment propositions in the forlorn hope of making a lot of money.

From those few documents that survive from the 1929–31 period, it would appear that his investments until then were fairly limited and not very productive. He still held over 12,000 STD 7 per cent preference shares and, nominally, nearly £12,000 of the STD 8 per cent guaranteed Notes. Another unproductive investment into which he was locked as a director of the company was Weymann's Motor Bodies. Out of a list of twenty-six investments held on his behalf by the National Provincial Bank, fourteen were either producing no dividend income or the companies had ceased trading. Despite this the bank was providing him with overdraft facilities against the security of these not very promising investments! With some of the money thus obtained he speculated in Brazilian Traction shares. In late 1929 he bought shares that he resold in March and April 1930. Although, from the incomplete figures available it is hard to see how he did not lose money on this investment, he must have been convinced that Brazilian Traction was still a good bet as he bought more shares later in the year. In June his stockbroker was writing that the balance sheet was strong and would 'make good in time' but a month later, in view of the financial crisis in Brazil, things were looking much worse. Coatalen continued to buy. By October Brazil's financial crisis had turned into a major political crisis, which led to the resignation of the President. The price of Brazilian Traction shares was down to about

---

13. Letters from Owen Clegg to Louis Coatalen (author's collection).

half the price they had been six months earlier, yet Coatalen still bought more shares. We do not know when he sold them, but it seems very likely that he lost quite a lot of money.

Paul Dormann

Another investment which sounds even more speculative involved a certain M. Paul Dormann who lived at 9 rue Michel-Ange, Paris XVIe but also had a villa in Biarritz.

It would appear that sometime in 1929 Louis Coatalen agreed to lend Dormann money to invest in the Banque Française de l'Afrique (BFA). Coatalen himself also bought shares in the same affair. Whether this arrangement was entered into before or after the Wall Street Crash in October is not clear but on 24 October Dormann produced a letter acknowledging his debt to Coatalen. In February 1930 Coatalen agreed with Lloyds & Provincial Foreign Bank that it would guarantee him overdraft facilities of Fr. 3,250,000 and that, in addition, it would provide Dormann with up to Fr. 2,500,000 on Coatalen's guarantee. The letter from Dormann, along with his mortgage documents, were deposited with National Provincial. At the end of November 1930 National Provincial was writing to express concern that Coatalen's own overdraft was up to Fr. 2,230,758 whilst that of Paul Dormann had reached Fr. 2,180,471, as 'you will realise that the Bank [BFA] shares have no market and their value is debatable'. They requested the advances be repaid or further suitable security be deposited with them.

Nothing seems to have happened, although Coatalen subsequently negotiated an overdraft facility in England. In March 1931 Dormann sent a letter to Coatalen to try to reassure him. He had apparently been in negotiation with Bauer Marchal, at the time a well-known merchant bank, to bail out the BFA but Bauer Marchal was itself in difficulty and could not help. He reported that he was happy to have found another partner, l'Union Commerciale & Industrielle of Paris (run by a financier called Schkaff), who envisaged a merger with the BFA and that this proposition was currently being submitted to experts from the Finance Ministry and the Banque de France. If all went well, he hoped that the deal would halve the amount of money he owed to Coatalen. At the same time Dormann said he was trying to sell his villa in Biarritz in order to pay back Coatalen but that the property market was extremely weak and he held out little hope of a speedy resolution. There was no longer any talk of making a profit, but he insisted he was doing everything in his power to minimise his losses.

Dormann's manoeuvring was to no avail, as the BFA ceased trading in August 1931. It emerged that Dormann had been the major shareholder in the bank and he was described as the *'maître occulte'* (hidden master) who had been pulling the strings behind the scenes, speculating in Romanian petrol interests. Not only were the tills empty but he himself was insolvent and the bank was finally liquidated at the end of the year. Coatalen must have lost out badly as a result.

The National Provincial Bank continued to express concern about the overdraft, as by that stage Coatalen owed them a total of

£44,000 (£27,000 in London and £17,000 in Paris) whilst a pencil note by Coatalen himself on the bottom of the letter suggests that the investments against which this was secured were optimistically valued at £42,000 (because of the large number of STD shares involved) and therefore no longer covered the overdraft.

It should be pointed out that the foregoing information is derived from surviving papers that relate to a relatively short period of Louis Coatalen's life and are probably not representative of his overall financial situation. All the indications are that by the end of World War I he was a wealthy man. Unlike W.O. Bentley, no trace has been found of him having received payment from the Government for his aero engine designs. One has to assume therefore that the source of his wealth came from his share in the profits of the Sunbeam Company. He lived as an affluent man for the rest of his life and still owned a number of properties in France at his death, despite these problems along the way.

Coatalen Compression Ignition Engines

Louis Coatalen's efforts to improve diesel engine performance were spread over more than a quarter of a century and were still preoccupying him when he was in his seventies, yet this aspect of his career remains one of the least known. Even though the work started during his time at Sunbeam and continued through his later career as an independent engineer, for clarity it is summarised here from start to finish.

Compression ignition engines were evidently important to him as he devoted a great deal of time and money to the subject, but the hoped for commercial results were never forthcoming. His interest in this area may well have resulted from his fascination with airships. By 1924 it was felt by the builders of these craft that, for safety reasons, it would be preferable to employ 'heavy oil' engines on future craft and this decision motivated much experimental work. The booklet published by Sunbeam to coincide with the 1929 International Aero Exhibition states that the development of the Sunbeam-Coatalen Compression Ignition Engine 'owes a great deal to the interest of the Air Ministry, and most of the Company's earlier work in connection with it was undertaken at the Ministry's request'.

The Experimental Department at the Sunbeam Works in Wolverhampton had started studying the subject as early as 1924 but it was in 1927 that a two-cylinder Benz engine was acquired for experimental purposes. The development work carried out over the following three years was summarised for Coatalen by A.G. Oates in a report that described the problems encountered and the progress made.[14] Oates wrote that the Benz engine (2 cyl. 135 x 200 mm) 'was of the automatic injection type with antechamber. Starting from cold or nearly cold was impossible without using the igniting cartridges in the cylinder heads, although the compression ratio was extremely high 17:1. Maximum power, at 800 rpm was 32 bhp (approx. 90 lbs/sq. inch bmep)'. It was then modified by fitting

---

14. A.G. Oates, *C.I.E. Report No. 13*, October 1930, typewritten thirty-one-page report (author's collection).

Chapter Seven

Top: Coatalen compression ignition engine, April 1930. This six-cylinder engine was based on the 'Dyak' aero-engine.

Bottom: Coatalen six-cylinder compression ignition engine on the test bed in January 1930.

and power output rose by 10 per cent. There then followed a number of experiments with lower compression ratios and different sizes of injection nozzle. Finally, having established that 13:1 was the minimum desirable compression ratio, aluminium pistons and lighter connecting rods were fitted and the engine speed was raised to 1,200 rpm (at 1,250 rpm it hit a very bad vibration period which prohibited going any faster). This Sunbeam-Benz engine then produced 60.6 bhp (114.5 lbs/sq inch bmep) – in other words roughly double the original power output.

In order to take developments further, the Experimental Department built their own single-cylinder (120 x 160 mm) engine which was finally run continuously for 160 hours (it had been intended to do a 200-hour trial but problems intervened) and was capable of turning at 2,000 rpm. It was used for numerous experiments with various injection nozzles in order to establish which provided the best penetration and detonation characteristics. As a young apprentice at Sunbeam in Wolverhampton, Norman Cliff was seconded to assist Alex Oates with these tests. He recalled that Oates 'was a man of engineering capability and possessor of a degree. He was a man of few words, giving no information to me as to the direction our experiments were tending. All data accumulated by us was carefully tabulated and filed away in a locked cabinet. He never failed to carry out his locking up ceremony when he went out of our cabin.'[15] Cliff described the single-cylinder engine as 'cunningly constructed both in the fuel pump

mechanically operated fuel valves with overhead camshafts, a constant output fuel pump with pressure reservoir, a new cylinder head with no antechamber, pistons with conical heads and 'nozzles of the directed-sprays type instead of the annular orifice type on the Benz injectors'. Cold starting was immediately infinitely better

---

15. Norman Cliff, *My Life at the Sunbeam*, 1990, p. 79.

and the cylinder head with a set of four gears in a swinging frame. This enabled two shafts to be moved relative to each other when the engine was in motion. By this means the lift of the needle injector valve could be increased or reduced and also the delivery from the pump could be controlled. By means of another control the pressure of fuel injection from the pump could be varied.'

By 1929, having got so far and obtained a great deal of knowledge and experience, it was decided to convert one of the Sunbeam-Coatalen Dyak engines to compression ignition operation. The Dyak, a six-cylinder 120 x 130 mm engine with an aluminium cylinder block, was the smallest of the range of aero-engines developed during World War I. Within three months of making the C.I. conversion Sunbeam were obtaining 139 bhp (102.5 bmep) at 2,000 rpm which compared well to the 100 hp at 1,200 rpm of the original petrol engine version. The engine would run smoothly right down to 200 rpm and the exhaust remained 'reasonably clear' up to 90 per cent of full load. The compression ratio used was 15:1. It was displayed at the 1929 Olympia Air Show and a report on the Show noted that the six-cylinder engine 'with its present weight can hardly be a feasible proposition for aeroplane use … it obviously has a long way to go yet. Solid injection is used at a pressure of 10,000 lbs per sq. in. and good atomisation is got by injecting direct onto the apex of the conically shaped piston.' A second engine with a cast iron cylinder block was also converted for 'transport' use and both were

Letter heading for Ets. L. Coatalen

then available for demonstration purposes. The main perceived advantage of using a diesel engine to power aeroplanes was that it would eliminate the use of a carburettor, a known weak point, and combined with the lack of an ignition system, this would greatly reduce the risk of fire which was also the key requisite for airships. Coatalen would no doubt have been somewhat frustrated at the time by the fact that by then the R 101 airship (built under the Imperial Airship Scheme and flown for the first time in December 1929) was committed to using Beardmore heavy oil engines. Admittedly the scale of the two engines was quite different, as the Beardmore Tornado straight-eight had a capacity of 84 litres against 8.8 litres of the Sunbeam, but in terms of comparable power output Coatalen's diesel engine produced 11.36 bhp per litre at 1,500 rpm (its maximum rose to 15.8 bhp per litre at 2,000 rpm) compared to the Beardmore's 7.73 bhp per litre. Crankshaft resonance problems of the latter meant that its maximum speed was reduced to 935 rpm at which it produced 650 bhp. At this time Sunbeam's Experimental Department had also produced a 24-litre, V-12, petrol engine, the 'Sikh III', which gave 1,000 hp at 1,650 rpm but

Chapter Seven

weighed 2,760lbs compared to the Beardmore's 4,773lbs. It is not known whether there were plans to produce a diesel version of this engine.

The poor financial situation of the STD Motors in the late 1920s has already been mentioned and car sales were very slow. It is not surprising that this aero-engine project was not a priority when the group was going through the first of its major financial restructuring exercises. Louis Coatalen, on the other hand, evidently thought he really was on to something important with his constant high-pressure injection pump and he purchased the diesel engine patent rights from Sunbeam for the sum of £10,500 despite the fact that he was well aware that the engine required a lot more development work. The first patent relating to compression ignition engines had been applied for in 1925 jointly with S.H. Attwood and together they were granted four further patents between 1928 and 1930. However, Coatalen must have come to an arrangement with Sunbeam to take over the project at the end of 1929 because in February 1930 five more patent applications were submitted in his name alone, which gave his address as Moorfield Works, Wolverhampton.

Even Owen Clegg, Managing Director of Automobiles Talbot France, advised him against the project, telling him that he was 'making the biggest mistake of your life' and in July 1930 was writing to Louis to:

> strongly advise you, even at this moment, to cut your losses on this particular job and not spend any more money on it. You have taken on experimental work which ought to have a big Concern with unlimited capital behind it if it is to be brought to a successful conclusion. You by yourself or with a small group can only hope to go on losing and losing until you have sunk the whole of your fortune in a problematic engine.[16]

However, Coatalen was not one to be daunted by a few technical or financial problems. Almost as soon as the diesel Dyak engine was running, he was working on ways of exploiting its future. In December 1929 Coatalen wrote from his address in Paris to Charlton Ogburn, a partner in Glenn, Alley & Geer, Counsellors at Law in New York, asking him to find a group of investors prepared to float a company to exploit the American rights to his patents. The engine was to be called the Sunbeam Coatalen High Speed Solid Injection Oil Engine. Although he admitted that the Sunbeam Company had not yet agreed to their name being used, he was hopeful that they would, in return for shares in the new company. He also wrote to the journal *Automotive Industries* in Philadelphia asking them to publish a statement to the effect that he had developed 'a high-speed solid injection oil engine of unusual performance characteristics' and that he was planning a trip to New York.

Early in February 1930 things started to move when Mr Laurence R. Wilder appeared on the

---

16. Owen Clegg, letter on Automobiles Talbot, Suresnes, headed notepaper dated 16 July 1930, to Louis Coatalen, Casa Tirrena, Capri (author's collection).

scene to negotiate on behalf of the Indian Motor Cycle Co. of Springfield, Mass. USA. Charlton Ogburn wrote to Coatalen to introduce 'my friend, Mr Wilder' whom he said was 'fully conversant with aviation matters. Several years ago he developed the Scintilla Magneto Company ... he founded the American Brown Boveri Corporation and for several years was the president of that corporation which owned also the New York Shipbuilding Company.'

Wilder visited England and went to the Sunbeam Works at Wolverhampton to inspect the engine. He cabled Coatalen, who was still in Paris: 'have seen engine and hereby accept offer ... to sell your patents present and future'. Coatalen was to receive £15,000 and a 5 per cent royalty on engines designed for them. The two men then met in London and agreed the basis of a deal. Wilder wrote a letter to Coatalen on Savoy Hotel, London, notepaper, dated 23 February 1930, 'confirming our understanding of today, will you please proceed to build a Coatalen heavy oil compression ignition engine of approximately 100 horsepower at 1,600 rpm of 750 lbs overall weight. My company will reimburse you with costs in connection with this work.'

Conveniently for Coatalen, just at this time *Silver Bullet* was being shipped to the USA for Kaye Don's attempt to set a new World Land Speed Record, so he was able to combine the journey to Daytona with a visit to New York to conclude the agreement with the Indian Motorcycle Co. Little did he know that it was going to be an extremely fraught and frustrating visit. He landed in America on 4 March and sailed home again on 16 April. Although he spent a couple of weeks in Daytona in the middle of this trip, trying to sort out the problems of the ill-fated *Silver Bullet*, most of the time was spent in New York.

The Sunbeam Motor Car Co. had evidently declined any involvement in an American diesel engine company as Coatalen, soon after arriving in New York, wrote to Keen Simons and Knauth, the public relations company retained to handle publicity for 'this remarkable Diesel Motor', stating that he 'would be obliged if you will see that the Sunbeam name is kept out when reference is made to my business transaction with the Indian Motorcycle Company. It must be clearly understood that my company is entirely out of this deal, which is a private one between myself and the Indian Motorcycle Company. As I explained to you on the telephone last night, I am very much upset and cross to see that, notwithstanding my expressed wishes on this subject, the name of my company has been used for advertising purposes and in reference to statements which are untrue.'

The thunder of the Coatalen-Indian deal was somewhat stolen by both Pratt & Whitney and Packard unveiling and even giving demonstration flights with aeroplanes powered by 'Diesel-type' engines at just this time. But Coatalen seems to have been unaware that the deal that Charlton Ogburn managed with great difficulty to strike on his behalf was actually just part of a much larger financial scam. Ogburn claimed that the package negotiated for the aviation rights alone was three times what Coatalen originally asked and included a shareholding in the company and a directorship. In the history of the Indian Company,

Chapter Seven

Harry Sucher wrote, 'it was suspected by a number of knowledgeable investors that the alleged dealings with Coatalen were undertaken to give favourable publicity to the company's doings in order to stimulate sales of the latest issue of common stock, all in the best "bull pool" tradition, and in no way were related to any *bona fide* plan to actually undertake aircraft engine manufacture.'[17] Whilst Coatalen was still in the US a lawsuit was filed that charged President Norman T. Bolles, Wilder and others with 'mismanagement and malfeasance in the conduct of the affairs of the company', and stated that certain officers and directors of the company had been able to purchase shares, sold to the public for $17 per share, at $5 per share. Unwittingly he found himself being used in a vortex of plots and counter plots.

Coatalen was back in Europe when on 6 May 1930 Charlton Ogburn wrote to inform him that 'Mr Paul Dupont (sic) of Wilmington was yesterday elected president of the Indian Motorcycle Company to replace Mr Bolles, who proved very disappointing and who yesterday resigned.' A couple of weeks later E. Paul du Pont himself cabled Coatalen: 'Think I should inform you as a director of Indian Company that I have been elected president of company and my brother vicepresident and other friendly interests have gone upon the board of directors. Messrs Bolles, Mitchell [and] Dodge have resigned. Am anxious to meet and become acquainted with you and look forward to your cooperation in developing our mutual interests under our royalty contract.' Larry Wilder appears to have continued to be involved with Indian for a while under the new ownership as he upset Coatalen by suggesting that the combination of the successful Packard diesel engine flight and the failure of *Silver Bullet* had undermined confidence in Coatalen's patents and engineering ability. He insisted that the only way for Coatalen to restore his reputation would be by an immediate successful test flight to prove the value of his design. This incensed Coatalen, not just because of the criticism of his ability but because under their agreement it was Wilder's responsibility to organise and finance this part of the deal.

On 12 July 1930 Paul du Pont replied to Coatalen: 'Mr Wilder's representations to the Board were to the effect that ... it would be advisable to have a flight in the airplane with your engine in order that the world can see the possibilities of it. As you are well aware, there is nothing so convincing concerning a piece of engineering than the actual performance in the field for which it is intended. Mr Wilder stated to the Board that he would bring about such a situation. It would be a great help to us if something of the sort could be done, and I sincerely hope Mr Wilder's attitude in the matter will not be construed as unfriendly.' However, it was clear that Indian Company had no intention of pursuing the matter, as the letter continues: 'Concerning the airplane Diesel engine, in the present state of depressed business, the matter has been laid in abeyance for the time being as problems have arisen in connection with the financial needs of the Indian Motorcycle Co. which make it inadvisable to attack this problem for the next

---

17. Harry V. Sucher, *The Iron Redskin*, Haynes, 1987, p. 210. I am grateful to David Burgess-Wise for drawing my attention to this extract.

few months.' Quite apart from the financial restraints, we also know that du Pont was by then concerned about the strength of the actual patents as he had been unable to gain access to them and a private detective had been employed to find out whether the Coatalen patents were any good and whether the transaction in which Wilder figured was straight or not. It is interesting to note that he reported the transaction was not straight and that the patents were no good, but that nothing could be done about it.[18]

This appears to have been the end of the matter as far as dealings with the United States were concerned, as no further correspondence from the period survives.[19] It is far from clear whether Coatalen made any money from this messy deal and there are indications that he was talked into putting up money that he would have had difficulty recovering. But far from being put off the project by the difficulties he had encountered, once he was back in France, he was soon energetically pursuing the next stage.

Already by the end of May Coatalen was preparing designs for a twelve-cylinder (140 x 150 mm) 600 hp aviation motor and had improved the design of the existing fuel valve. He had had meetings with the technical director of Lorraine Dietrich who was very interested in the concept but Coatalen had turned down cooperation, as he realised that his motor would just be absorbed and become another Lorraine motor. He explained to du Pont, 'I am sufficiently well off to wish to keep for myself the child that I have brought off in this world.'

In June he had started to set up a business called Moteurs Coatalen, initially as a development facility but evidently with the longer-term expectation that aero-engine manufacture might be started. A workshop was set aside within the premises of Avions Weymann-Lepère at Levallois-Perret and a budget was prepared that set out how Moteurs Coatalen would contribute to Weymann's overheads with Louis injecting the initial working capital of Fr. 62,000.

Charles Weymann is best known for his patented bodywork system for automobiles that he developed based on his experience as a World War I pilot.[20] Coatalen and he appear to have been genuinely friendly over a number of years, and when Weymann decided to set up a company in England to manufacture car bodies Louis Coatalen probably helped to finance the business and became a director. Weymann's Motor Bodies (1925) Ltd had its registered offices at 22, Austin Friars in the City of London with the works, firstly in Putney and then from 1928 in Addlestone, Surrey. Weymann himself was Chairman whilst Edwin Izod was Managing Director. The other directors were: Lt Col Llewelyn Evans, Jessie Frank and Maurice Dollfus. A letter from Hermann Aron (of Rotax) survives from which it is clear that Aron had also lent money to Weymann. Aron wrote to Coatalen in July

---

18. Information contained in a letter dated 12 June 1930 from the du Pont archive, provided by Hayden R. Shepley.
19. Most of the information comes from a file of correspondence kept by Louis Coatalen now in the author's care.
20. Charles Terres Weymann patented his flexible bodywork system in France in 1921 and started manufacturing at Levallois. He set up a company in England to license coachbuilders to use those patents in 1923 before setting up a factory in Putney two years later. He had learnt to fly before World War I in a Santos-Dumont plane and during the conflict he flew with the French air force.

Chapter Seven

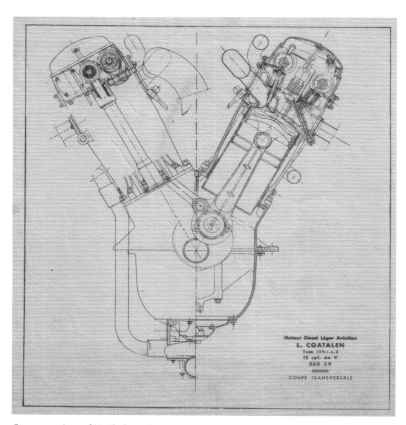
Cross-section of V-12 Coatalen compression ignition engine.

1930 pointing out that Weymann still owed him £3,000 and remarking, 'when we lent him money before, you got all yours two years before I did mine'.

It would seem that soon after agreeing to set up this aero-engine facility, Coatalen's health took a turn for the worse and he retreated to the island of Capri, leaving Charles Weymann in charge. The converted Dyak engines, the single-cylinder experimental engine and the engineer Alex Oates arrived from England and by the first week of July everything was in place so that Weymann was able to invite the celebrated journalist Charles Faroux to witness a test run. Weymann reported to Coatalen that Faroux had been suitably impressed by how well the Dyak ran, by the cleanliness of its exhaust and how quickly it could be accelerated. Within weeks, engineers from Skoda were negotiating to purchase an engine and the manufacturing rights, although Oates, assisted by Bridson and Monnier, encountered a number of problems in keeping the Dyak operational and so the appointments by Skoda to inspect the engine had to be carefully managed. Another important visitor to inspect the engine and express an interest in it was André Mariage, Director of Société des Transports en Commun de la Région Parisienne. Throughout the summer and autumn of that year reports and telegrams were exchanged between Paris and Capri but despite the importance of some of

Collapse and Recovery 1930–39

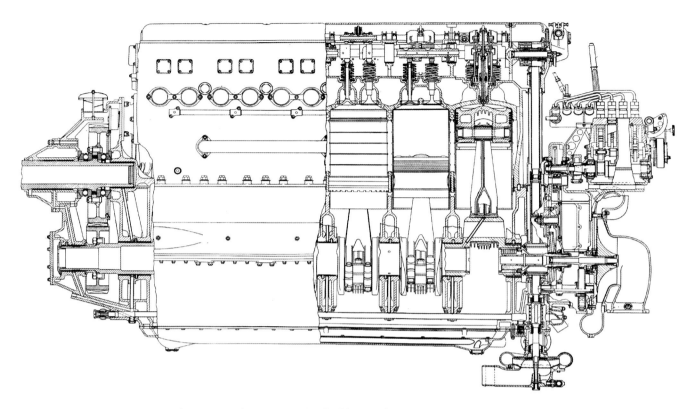

General arrangement drawing of V-12 Coatalen compression ignition engine.

the negotiations and major issues arising in his other business interests that demanded his presence, Coatalen remained in Capri throughout. One senses the somewhat erratic nature to his communications and at one point a secretary wrote on his behalf that Coatalen was suffering so badly that he could not write himself. However, one piece of good news came through at the beginning of October when a letter was received from the Service Technique de l'Aéronautique of the French Ministry of Defence confirming that they were interested in the *'moteur Coatalen à huile lourde'* and that it would be presented to the next CEPANA (Commission d'Examens Permanent de Projets d'Appareils Nouveaux pour l'Aéronautique).

There seems to have been a pause in diesel engine development work whilst Coatalen remained in Capri throughout the winter until spring 1931. When he had recovered, a six-cylinder Panhard commercial vehicle engine was modified in 1933. This was used as a test rig to demonstrate the superiority of his mechanically controlled injection system which delivered fuel during the whole of the combustion cycle. The power of the engine increased from 100 cv to 130 cv and it could be slowed right down to 130 rpm or accelerated up to 2,000 rpm with an entirely clean exhaust. That Coatalen was really back on form was confirmed by Maurice Victor, who wrote of his visit to see this engine towards the end of 1935. '*M. Coatalen m'a*

Chapter Seven

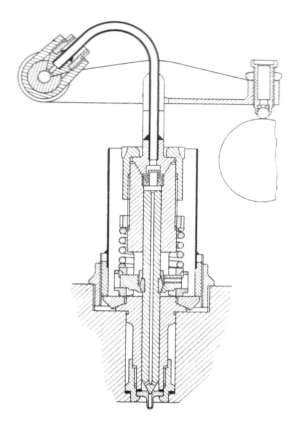
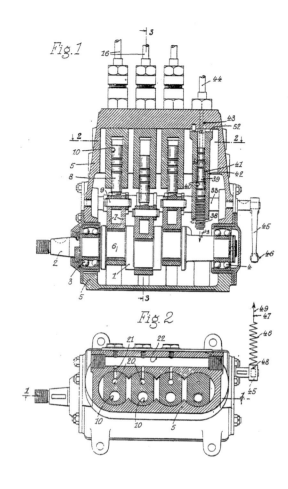

Left: Fuel injection unit.
Right: Fuel injection pump.

épaté. On le sent enthousiaste et méthodique … un moteur à huile lourde qui marche mieux q'un moteur à essence'[21] (Mr Coatalen astounded me. He is enthusiastic and methodical … a heavy oil engine which runs better than a petrol engine).

Although designs for a twelve-cylinder engine had been on the drawing board in 1930, these were never realised and it was not until 1935 that Coatalen succeeded in creating a V12 compression ignition engine when he acquired four Hispano Suiza engines from a crashed aeroplane and two were converted to operate with his patented constant pressure injection pump.[22] These 150 mm bore x 170 mm stroke motors had a capacity of 36,050 cc and employed a centrifugal fan-type supercharging blower that ran at ten times crankshaft speed and could maintain atmospheric pressure up to 9,800 ft. Although the basis of the engine was Hispano, it underwent a major re-design and re-build to become the Coatalen V-12. The original cylinder blocks incorporated single overhead camshafts operating two vertical valves per cylinder and had twin horizontal sparking plugs.[23] Coatalen replaced these by a new cylinder head design that featured four

---

21. Maurice Victor, *Les Ailes*, 5 December 1935.
22. Hervé Coatalen was involved in helping to recover the engines from the crash site.
23. I am grateful to Karl Ludvigsen for providing details of the Hispano engine construction.

inclined valves per cylinder, actuated by pivoted intermediary fingers, driven by twin overhead camshafts on either side of the fuel injectors. The latter were placed directly above the centre of the piston and sprayed fuel radially at 10,000 lb per sq in through ten 0.13 mm diameter jets. The blocks were cast in 2L5 alloy with hardened steel liners. New domed crown pistons and connecting rods designed to cope with the higher compression ratio of 14:1 were fitted. To start with, articulated connecting rods (master and link rod) – similar to Coatalen's World War I aero-engines – were used, as the Hispano Y-type engine used this system but by 1938 the crankshaft had been redesigned with 'fork and blade' connecting rods. Forced lubrication was even provided to the gudgeon pins. Two fuel pressure pumps were mounted above the blower and the fuel injection worked on the common rail system with mechanically operated injection valves. Injection was timed to commence 25 degrees before TDC. The engine gave 550 hp at 2,000 rpm and had a dry weight of 1,210 lb.

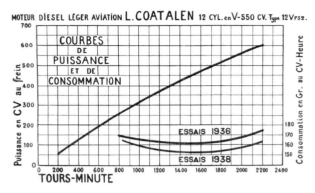

1938 Power curve for the V-12 diesel engine showing output of 600 bhp at 2200 rpm and consumption reduced to 154 gr per horse power hour.

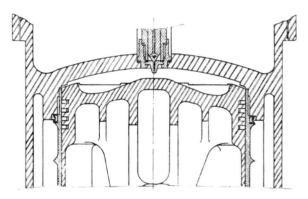

Drawing of cylinder head design for the diesel engine which was cited as exemplary.

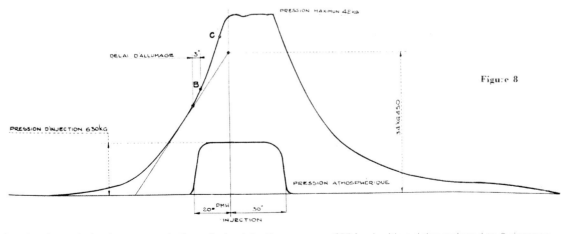

Diagram showing the variation in pressure in the cylinder. Injection pressure 630 kg, ignition delay reduced to 3 degrees and combustion pressure reduced to 42 kg.

Chapter Seven

V-12 Coatalen compression ignition engine on the test bed.

Coatalen displayed his engine at the 1936 Paris Air Show and in 1937 it successfully passed its preliminary official tests to satisfy the French Air Ministry, which was actively seeking to foster the development of heavy-oil engines and had earmarked Fr. 10 million for that purpose. He was also granted patents for an improved version of his fuel injection pump, his cylinder head design and other improvements to fuel injectors. A full description of the engine was published in *The Automobile Engineer* to coincide with the 1938 Paris Air Show. It was reported that development of the engine was continuing but 'as a result of his researches M. Coatalen is confirmed in his opinion that maximum cylinder pressures must be kept reasonably low, and to avoid undue pressure rise at the commencement of the injection period, a constant-pressure injection system must be employed'.[24]

Louis Coatalen's interest in diesel engine development for use in aircraft was perhaps typical of his approach to engineering. He would have been aware that it was a subject that was interesting leading engineers of his time and worked hard at trying to make his own distinctive contribution to the subject. He was certainly not alone in pursuing the subject and it is notable that others with far greater facilities at their disposal did not succeed in turning the diesel engine into a commercial aeronautical proposition any more than he did.[25] However, he was undoubtedly at the forefront of developments at the time. In England in the 1920s the Royal Aircraft Establishment at Farnborough conducted research into diesel engines that was published in 1927 and this was crucial in motivating Coatalen to seek

The completed V-12 engine prepared for display.

---
24. The Coatalen Diesel, *The Automobile Engineer*, March 1938, pp. 77-81.
25. Sir Harry Ricardo, *Memories and Machines*, Constable, 1968, p. 240: 'our attempts to develop a diesel engine light enough for aircraft propulsion though it taught us much showed that such an engine could not compete with a comparable petrol engine running on the high octane fuel available for aircraft by the middle 30's (sic)'.

improvements. It is worth remembering that although Rudolf Diesel's original patent dates back to 1892, it was only in the late 1920s that it became a viable power unit for road transport use. Diesel-powered lorries and cars are now so ubiquitous that we might not be aware that in Britain in 1928 they were 'an object of curiosity'. At the 1929 Commercial Motor Exhibition (the year the first Mercedes Benz bus with a diesel engine was brought to England and put into service) there were only three makes showing oil-engined vehicles and these were all foreign. By the 1931 Exhibition things had begun to change and there were fourteen British makes offering diesel-engined commercial vehicles.[26] However, the first commercial introduction of diesel engines for use in private cars did not occur until several years later.

That Coatalen was not alone in pursuing the objective of using a compression-ignition engine for aircraft can be illustrated by a few international examples. For instance, in America in 1931 the Packard diesel engine already mentioned broke the world record for endurance flying without refuelling, remaining in the air for eighty-four and a half hours. That year in Italy a six-cylinder Fiat engine converted to diesel operation with six fuel pumps, flew from Turin to Rome. In 1933 a British Bristol 'Phoenix' diesel engine was flown and subsequently set up the world's altitude record for that type of engine at 28,000 feet, thus demonstrating convincingly that power at altitude decreases more slowly with a diesel engine than it does with a gasoline engine.

During his official visit to the 1936 Salon de l'Aéronautique, the President of the French Republic, M. Lebrun, congratulates Louis Coatalen on his V-12 Diesel engine.

Coatalen's contribution to the subject matter was recognised, for example, in an article written in 1949 by Jean Aury that reviewed the influence of combustion chamber shape on diesel engine performance. His combustion chamber was cited as a model of its type and its consumption was said to have set new records.[27] According to Charles Faroux, although his engines never went into production in France, the Russians equipped all their large tanks with engines based on Coatalen's patents without ever paying a rouble![28] It is clear that it was a subject that continued to fascinate him, as in March 1955 when he was President of the Société des Ingénieurs de l'Automobile he wrote an article for the SIA Journal summarising his achievements. He was then aged seventy-six.

---

26. The Editor of The Commercial Motor, *Compression Ignition Engines for Road Vehicles*, Temple Press, 1932.
27. Jean Aury, 'Influence des formes des chambres de combustion sur les performances des moteurs Diesel', *Journal de la S.I.A.*, April 1949.
28. Charles Faroux, 'Allocution', *Journal de la S.I.A.*, July 1953.

Chapter Seven

Reconstruction and Recovery

Even during his year in Capri, Coatalen could not remain idle for long and it appears that before he went to the island he was putting in place plans for life 'post-Sunbeam'. The aero-engine project has been described, for which the Moteurs Coatalen business was set up; this later became Ets. L. Coatalen SA, with offices at 27 rue Jules Verne, St Ouen. He had also invested in la Société Francaise des Freins Hydrauliques Lockheed that was to be his main centre of activity for the next thirty years.

It seems that negotiations had also commenced for him to become Technical Adviser to the STD Group. The timing of this is unclear but it is likely that this role was partly a reflection of his declining health in 1930 but also a desire to free himself from the constrictions of the Group. He was to provide his own offices and was to be paid £3,000 per year (as Managing Director he had earned £4,000 p.a. plus 3 per cent commission on Sunbeam profits) but the company would pay the salary of two 'competent draughtsmen'. This proposal was a step towards him becoming non-executive but only a draft of the agreement survives. For a while, at least, he did set up a design office in Capri so he continued to work intermittently throughout his illness. The new Chairman, Sir Travers Clarke, corresponded with Coatalen in the months that followed the appointment of new directors about trying 'to fix up your agreement with STD Motors as quickly as is possible' and saying that he 'quite understood your desire to work in peace and quiet in the same way as Royce does'. This is a reference to Sir Henry Royce who had been advised by doctors to avoid Derby and to live by the sea because of his frail health. For many years, aided by a team of draughtsmen in the south of England and the South of France, Royce developed new designs without going near the Rolls-Royce works. Coatalen evidently imagined a similar situation for himself. Travers Clarke urged Louis to meet with him personally on one of the directors' three weekly visits to Automobiles Talbot in Paris to discuss the heads of agreement. An on-going role for Louis was envisaged but nothing appears to have come of it. Whether the directors got tired of waiting for Louis to return from Capri, or they were unable to agree terms, or Louis decided he was no longer interested, is not known.[29]

Lockheed Brakes

Between 1917 and 1923 the American Malcolm Loughead (tired of having his name mispronounced he later changed it to Lockheed to match its correct pronunciation) was granted seven patents for his hydraulic four-wheel brake system and it was this system that was developed in America for use on production cars. The Chrysler Six was launched in 1924 equipped with 'Chrysler Lockheed Hydraulic Four-Wheel Brakes'. In the same year the Triumph 13/35 was the first British car to

---

29. Typewritten letter from Travers Clarke, 12 Princes Street, Hanover Square, London, dated 12 May 1931, to L. Coatalen, Casa Tirrena, Capri.

Collapse and Recovery 1930–39

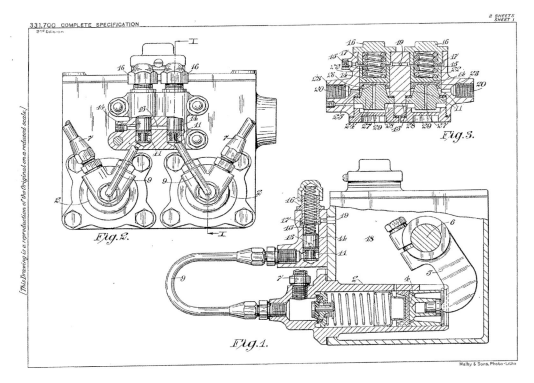

Patent for dual master cylinders for hydraulic brakes applied for in July 1929 and granted a year later.

be fitted with Lockheed hydraulic brakes. It was Gustave Baehr, well-known Parisian distributor of motor cars, who bought the exclusive licence for France and set up la Société Française des Freins Hydrauliques Lockheed in 1928. In May 1930 Louis Coatalen became involved financially with the French company and by August the first hydraulic brakes were delivered to Automobiles Talbot for testing. However, a search of his patent applications reveals that he had already proposed three hydraulic brake improvements in the UK in July 1929. Two of these patents were granted (covering dual master cylinders and a means of preventing oil loss should a pipe leak) whilst the third would have been granted but the required fee was not paid. In 1931 he seems to have taken direct control of the firm. It appears that while he was still living in Capri he involved Henri Perrot in the business as his representative on site. From a letter to Baehr, after the latter had resigned, it is clear that Henri Perrot had been the intermediary through whom Coatalen had indicated that he would put up the necessary funds. Together Coatalen and Perrot built it into a successful business as the demand for hydraulic brakes expanded. They evidently also sought other openings for the hydraulic systems they manufactured as Coatalen was granted patents in France for a hydraulic gear changing apparatus for automatic gearboxes, for hydraulic controls of air brakes on aircraft, for hydraulic carburettor adjustment, and for a hydraulic or pneumatic telescopic device for controlling aircraft undercarriages before World War II.

Chapter Seven

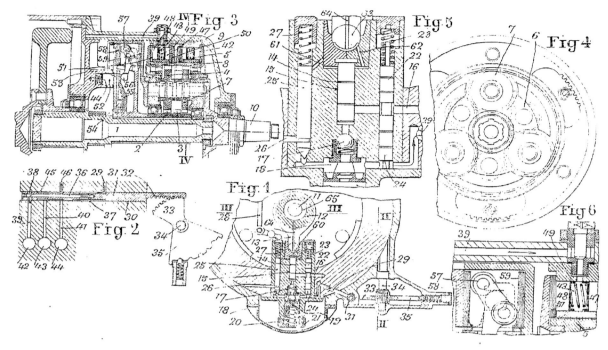

Patent for automatic gearbox with hydraulic control granted in 1935.

K.L.G. Sparking Plugs

We know that the links between Louis Coatalen and Kenelm Lee Guinness stretch back to the earliest days of their careers before World War I when they were both young racing drivers. Guinness started to manufacture sparking plugs for racing cars in 1912 and applied for a patent in 1913, but with the beginning of the war he extended production to provide plugs for military purposes. The business grew substantially so that in 1916 it was decided to form a limited liability company, the Robinhood Engineering Works Ltd. The first board of directors consisted of Louis Coatalen, Kenelm E. Lee Guinness, Harold Percy Hawthorn and Noel van Raalte. In 1919 an arrangement was reached with S. Smith & Sons (Motor Accessories) Ltd in which Smiths became the sole agents for the sale of K.L.G. Sparking Plugs. Guinness finally agreed to sell out in 1927 so Smiths acquired the whole share capital of Robinhood Engineering Works and thus also took over the manufacturing side of the business as well, renaming it K.L.G. Sparking Plugs Ltd. At that period mica insulators were used and although experiments with ceramic insulators were undertaken at the end of the decade these were not successful. It was not until 1935 that Smiths entered into an exclusive agreement with Bosch to manufacture ceramic sparking plugs in Great Britain under licence. 'As a result of the agreement, K.L.G. obtained from Bosch a secret formula for the manufacture of ceramic insulators with fused alumina,'[30] which were marketed under the

---

30. Monopolies & Mergers Commission Report, *S. Smith & Sons Ltd, History and Development 1900–1930*, p. 91.

Collapse and Recovery 1930–39

trade mark 'Corundite'. In return Bosch was assigned one tenth of K.L.G.'s shares. During World War II K.L.G. made considerable technical progress in the use of new materials and after the war was able to replace Corundite with a new improved ceramic material called 'Hylumina'. In 1944 it cancelled its agreement with Bosch and re-purchased the shareholding.

Coatalen's involvement as a director of K.L.G. in England had not lasted very long, as he resigned in 1918, but his personal links to 'Bill' Guinness remained close and Sunbeam racing cars continued to use K.L.G. plugs. Smith's arrangements for marketing K.L.G. plugs in France were initially handled by a subsidiary company, Smith Kirby, until 1934 when that firm was wound up. In its place, a new K.L.G. company was set up in France that 'was two-thirds owned by Smiths, the remainder by longstanding French partners.'[31] It seems probable that Coatalen was one of those partners as, five years later in 1939 he took over the complete ownership of the Bougies K.L.G. business in France, becoming the 'administrateur délégué' (Managing Director) with Gustave Delage as 'president du conseil' (Chairman). The works were in premises in Suresnes that had previously been used by Talbot for repairs, but he soon set about seeking a new factory for spark plug production.[32] It seems that through contacts of Henri Berthier, future father-in-law of his son Jean, he was made aware of the unused Provot hat factory in Chazelles, near Lyon, and set about purchasing it. It would appear that production had not been started by the time war broke out as it was then transformed temporarily into a barracks for a battalion of reservists.[33]

This joint K.L.G./Lockheed advertisement appeared in the programme for the 1939 La Baule Grand Prix which was never held due to the outbreak of war.

---

31. James Nye, *A Long Time in Making, the history of Smiths*, Oxford University Press, 2014, p. 90.
32. *Ouest Eclair*, La Société des Bougies KLG devient entièrement française, 25 January 1939.
33. Pierre Mathieu www.patrimoineethistoiredechazellessurlyon.fr

Chapter Seven

A 1937 advertisement for KLG spark plugs introducing platinum electrodes.

Private Life

It is believed that Louis had started having an affair with Ellen Mary Bridson (1896–1973), who was always known as 'Dickie', towards the end of World War I (during which she drove an ambulance), as she started to keep a scrapbook about Sunbeams and Coatalen at that time. She was apparently 'a good Catholic girl from Oxford' but must have remained in love with the charming Frenchman, as it was she who picked him up after Iris, his third wife, left. 'Dickie' was the one responsible for taking Louis away to Casa Tirrena on the island of Capri during the summer of 1930 in order to remove him from access to drugs and alcohol, and eventually partially restored his health. According to Marjolie Coatalen this was not entirely straightforward, as he would get the crew to stock up on duty-free gin and sneak down to the yacht to drink, out of Dickie's sight. Apparently, she had had two abortions before she and Louis were finally able to marry in 1935, after eighteen years of waiting. There were no children from their marriage, but they remained together until he died.[1]

Thanks to the ministrations of Dickie on Capri, Coatalen returned to Paris in 1931. He was soon back at work on his diesel engine projects and at Lockheed. A few years later in October 1934 (which curiously coincided with negotiations for STD Motors to be finally taken over by Rootes Securities but, there is no evidence that Coatalen was involved) he was one of a party of members of the French Société des Ingénieurs de l'Automobile that travelled to London to attend the fifteenth annual dinner of the English Institute of Automobile Engineers held at Park Lane Hotel, Piccadilly, London. The French group was led by SIA president Maurice Goudard (Solex) and included, amongst others, Coatalen's friend and colleague Henri Perrot and Maurice Rollo (General

'Dickie' Bridson with her dog. She was Louis Coatalen's fourth wife.

---

1. Marjolie Coatalen, unpublished autobiography draft, pp. 6 and 21.

Chapter Seven

Louis and Dickie returned to Capri regularly for holidays during the 1930s. They were often joined by both his sons, Hervé and Jean, and his daughter Marjolie. Here he is seen crossing the piazza in 1934, accompanied by his younger son Jean on their way to lunch. In 1951 he gave Casa Tirrena to Marjolie.

Motors, France). Laurence Pomeroy, as President of the IAE, was in the Chair and in humorous mood. He referred in his speech to the development of the diesel engine and suggested that 'in this matter the industry was to some extent like Columbus who, when he discovered America, did not know where he was going, when he arrived he did not know where he was, and when he got back home did not know where he had been'. One wonders if these remarks were partly aimed at his old friend Coatalen.

Coatalen's divorce from Iris did not come through until April 1935 and so he and Dickie were married in Paris in November that year. However, financial wrangling with Iris must have continued despite the divorce, as finally the Parisian Court of Appeal ruled in March 1939 that Coatalen had to pay to Iris the sum of Fr. 420,000 in settlement of their pre-nuptial agreement. Coatalen had argued that it was a sum he had put up in 1918 to reassure Iris of his intention to marry her when he was free to do so and that therefore, having married her, he had carried out his obligation and owed nothing more. He felt he was free to use the money as he saw fit. The judges did not agree.

Iris and her partner Yvonne were evidently able to live comfortably at this period as they owned an early Rolls-Bentley

Collapse and Recovery 1930–39

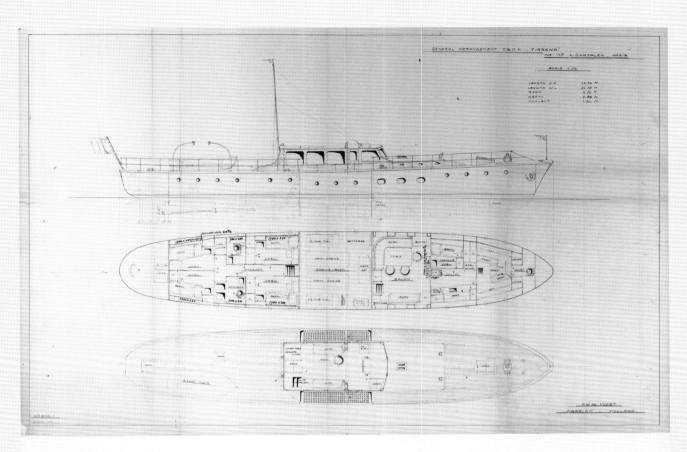

Top: H.W. de Voogt's original general arrangement plans for *Tirrena*, 1937.
Bottom: Tirrena undergoing trials in Holland before delivery, 1938.

Chapter Seven

and a 4¼ Bentley at different times in the years leading up to World War II.² When the Germans invaded in June 1940 and Iris escaped to England, she was devastated because she not only had to abandon her Bentley but also her prized Purdey rifle.³

Louis' financial situation must have been greatly improved as in October 1937 he commissioned a new yacht to be built at the De Vries boatyard at Aalsmeer in Holland to the designs of the naval architect H.W. de Voogt. In addition to cruising in the Mediterranean, it was intended also for use on France's inland waterways. This motor yacht, *Tirrena*, 26.5-m long, was powered by two Gardner diesel engines. It was comfortably fitted out with a row of three, very distinctive, large oval portholes on each side of the saloon and had four spacious cabins in addition to the captain's cabin. She was launched in May 1938 and the boatyard still has a film of the yacht undergoing trials and then of Louis and Dickie taking delivery. Under Captain Bruhier it undertook its first long voyage to Algiers during that year. Sadly, it only had a short career, as in 1940 the Italians sank it in order to block the entrance to a canal at Livorno.

Tirrena undergoing trials in Holland before delivery, 1938.

2. 1933 Rolls-Bentley B5 AE, ex-London show car with Vanden Plas body; 1936 Bentley 4.25 with Vanvooren cabriolet body; 1947 Bentley Mk VI standard steel body (Will Morrison).
3. More traumatically, Iris was also separated from Yvonne but they were reunited after the war and lived together until Yvonne's death.

Eight

The 1940s and After 1940 – 62

THE political storm clouds that led to the outbreak of World War II in 1939 had been building up for a while. After the Anschluss of Austria, France had called up its 400,000 reservists in 1938. However, most of the population was completely unprepared for the speed of the German invasion when it came in 1940. During World War I German forces had not reached Paris and since then France had invested heavily in the defence of its Eastern border by constructing defences known as the Maginot Line, so they perhaps did not feel under immediate threat. It was a shock when Germany invaded Belgium, Luxembourg and Holland on 10 May 1940 and swept right through to Paris in little over a month, entering the French capital on 14 June. Suddenly the French in the northern part of the country were having to come to terms with living under enemy occupation.

The Germans rapidly took control of the automobile industry to use it for their own purposes. Businesses which did not quickly re-start production soon found their machinery being uplifted and transported across the Rhine for use in German factories. Entrepreneurs were therefore faced with a choice between losing their investments entirely or cooperating reluctantly in order to maintain their equipment and to provide their workforce with continued employment. As the war progressed the problem of keeping workers gainfully employed became more and more important, for without it they risked being deported to Germany under the forced labour scheme. There was no alternative to supplying the enemy because this is what they demanded, but it was also the only sure way of obtaining raw materials and payment for the orders placed. Subsequently it became a question of trying to hide stocks and materials from the occupants, delaying

Chapter Eight

deliveries on orders received or supplying sub-standard products where possible.[1]

This was no doubt the situation that confronted Louis Coatalen who was sixty years old at the beginning of the war and comfortably installed in Paris. Despite his British wife and the fact he, too, was naturalised British, he remained in Paris throughout, keeping his business ticking over. It seems that there was a plan to move the works to Libourne, near Bordeaux, ahead of the invasion but no machinery was actually moved. During the conflict there was evidently a resistance movement within the K.L.G. factory, as a clandestine newspaper, *Le Travailleur*, was published, which protested against deportation, etc.

Louis' granddaughter, Carole de Chabot, who lived with Dickie and him through much of the war as a very young child, has written her memoires, which give a good idea of their life at that time. They had a large second floor apartment at 7 rue Lesueur, Paris 16eme, which she recalls had long, dark corridors. In addition to Carole's nanny, Luce Dartevelle, known as 'Nounou', Henriette the cook, a butler, Emile the chauffeur, and a laundry lady were employed; the flat was also inhabited by two dogs, an enormous Berger de Brie sheepdog and a tiny Pekingese. The butler, who served at table, wearing a special yellow and black striped gilet and white gloves, would be summoned by Dickie ringing a bell when it was time to clear or serve the next course.

They also had a house in the Yvelines countryside, south-west of Paris, some 15 km from Versailles called Le Chêne, in Les Mousseaux, near the village of Jouars-Pontchartrain. Here vegetables were grown, including tomatoes, asparagus and strawberries, and there was also a cow so they had a good supply of milk and cream. Even during those difficult years they were able to live comfortably.

Coatalen's two sons managed to get to England and join the Allied forces. His elder son, Hervé, who had been called up to do military service in France in 1937, escaped aboard a Dutch freighter from the mouth of the river Gironde in June 1940 shortly before the Germans got there. He then joined the RNVR, initially at Felixstowe and Dover as a Sub-Lieutenant Engineer Officer for the 4th Flotilla of motor torpedo boats. At the end of 1942 he was posted to Plymouth as Base Engineer Officer at HMS *Defence* and it was there that he met his future wife, Anna. Then in September 1943 he was posted to the Mediterranean, firstly in Malta and afterwards in Sardinia and Corsica, maintaining the operational flotillas of Motor Torpedo Boats for the Advanced Coastal Forces. Near the end of the war, following a close encounter with a bomb, he had ear trouble and was given a shore job in Hatfield with de Havilland, working on the prototype Sea Mosquito.

Louis Coatalen's younger son, Jean, became a pilot with the Fighting French Air Force.

Jean Coatalen flew a Spitfire for 340 (FF) Squadron based at Hornchurch in Essex under the pseudonym, Joseph Lambert. He was

---

1. For information about this period, see: Renaud de Rochebrune and Jean-Claude Hazera, *Les Patrons sous l'Occupation*, Odile Jacob, Paris, 1995 re-printed 2013.

extraordinarily lucky to survive, as he is known to have had at least three, and probably four, crashes. The first was while training as a pilot in France before the invasion. The second was during a mission escorting fighter-bombers to raid St Omer airfield, when he was shot down in flames on 30 July 1942, landing in occupied territory. He parachuted to safety and was given civilian clothes and food by some locals at Serques before making his way to Paris. When he got to his father's flat nobody was there but the concierge let him in. Louis came back the next morning, having received a message that Jean had arrived. He took him to the doctor to treat the burns to his arm sustained during the flight. Two days later Jean left to try to get to the non-occupied zone in the southern part of France. After a number of attempts he managed to re-join his wife and parents-in-law near Lyon but the authorities became aware of his presence and he 'was induced under threat of imprisonment by the Vichy government and by fear of reprisals on my family to give my parole not to leave the district'. When the Germans occupied the whole of France in November he considered that his parole automatically came to an end and began to seek ways of escaping to England. It was not until August 1943 that a route was organised and he made his way across France to the Pyrenees, arriving in Andorra five days later. This was followed by a week's walk to Manresa, from where a train was taken to Barcelona. There he reported to the British Consulate and was sent on to Madrid and then to Gibraltar, finally

Jean and Hervé Coatalen in uniform. Jean flew with the RAF and Hervé was with the RNVR.

arriving back in the UK on 26 September 1943.[2] He was soon back in the air and flew a total of 116 missions for which he was awarded the Croix de Guerre, the Médaille des Evadés, and later promoted Officier de la Légion d'Honneur. According to the recollections of his half-sister Marjolie, he was shot down on another occasion when he was able to escape by getting to Marseille, crossing to Algeria and then across the desert by camel to Morocco, but no verification of this incident has been found.[3] The final accident was after the Allied landings, when taking off in his Spitfire from

---

2. National Archives, *Most Secret Report 'Evaded Capture in France'*, WO/208/3315.
3. Marjolie Coatalen, cited in Carole de Chabot, *Rebelle avec l'aide de Dieu*, memorizon, 2012, p. 132. No corroboration of this story has been found.

Chapter Eight

the Sommervieu airfield in Normandy to undertake armed reconnaissance, his engine exploded and caught fire and he was wounded and taken to hospital.[4]

Louis' elder son, Hervé, had married Betty in 1939 and their daughter Carole was born in February 1940. However, when they escaped ahead of the German invasion, Carole was left in the care of her 'nounou' Luce. Luce brought Carole up with her own family until Louis discovered where his granddaughter was and brought them both back to live with him and Dickie in Paris. They would alternate between the flat in Paris and the house at Les Mousseaux, which was cared for by a gardener and his wife who lived in a neighbouring house. This couple rescued and hid three British airmen who had been shot down in the area. They burnt the uniforms and passed the airmen on to other members of the Resistance to escape. The next morning the Germans arrived to search the place and found one button that had not been destroyed. The couple were arrested and deported to a prisoner of war camp. Louis and his wife were at Les Mousseaux at the time but managed to scatter and to get away across country before the Germans came to search and make enquiries. Under the pressures of living and working under the occupation, Louis once more succumbed to taking drugs.[5] He only took out one patent during World War II, which was for the design of an apparatus to compress gas. The gardener and his wife survived the war but when Louis saw their physical condition on their return, he was reduced to tears.[6]

After the war, when they were demobilised, both sons returned to France as prospects appeared brighter there than in England where most things continued to be tightly rationed. Jean, after working briefly for his father, worked for Duckhams Oils at Levallois-Perret while Hervé, who had just remarried (his first marriage had not survived the separation of the war), joined

K.L.G. Corundite Spark Plug advertisement.

---

4. Information from Daniel Carville, www.francecrashes39-45.net 17 September 2017.
5. Marjolie Coatalen, cited in *Rebelle avec l'aide de Dieu*, p. 120.
6. Carole de Chabot, *Rebelle avec l'aide de Dieu*, p. 15.

his father in the K.L.G. Sparking Plug business, having been promised a job and a honeymoon. He never got the honeymoon and the job was far from a bed of roses!

In 1946 Hervé was set the task of getting the derelict factory in Chazelles-sur-Lyon, which K.L.G. had purchased just before the war, fitted out and into action. Helped by a grant from the French Ministry of Aviation it was to manufacture ceramic insulators instead of the mica insulators that K.L.G. had used previously. Hervé was promised a directorship once this task was completed with a renewable five-year contract. This he was duly granted but working for Louis turned out to be fraught with difficulties. One was liable to be fired for arriving a little late for a meeting. In 1949 business was apparently slowly improving with orders from Renault for 100,000 isolators and even a small order for K.L.G. plugs from Mr Sletter for Rolls-Royce cars in France. By then Hervé was Managing Director, working at the main factory at Saint Cloud but making regular trips back to Chazelles to keep an eye on the works there. Finances were tight and meetings were held to explore which Lockheed parts could be machined at the K.L.G. works. Louis Coatalen remained President of the firm and, during a holiday on the outskirts of Concarneau in Brittany, fired off regular letters to his son and to Henri Perrot whom he entrusted to investigate a difficult cash-flow situation. Once he got back from holiday, he continued to criticise his son's management of the business and tried to demote him, suggesting he should just run the factory at Chazelles.

How this was resolved is not clear but

For the first post-war Monte Carlo Rallye held in January 1949, the Coatalens had a caravan at the Paris start that had been converted as a 'hospitality unit' with a signboard advertising the K.L.G./Lockheed Bar. Louis Coatalen stands outside (left) while his son Hervé can just be seen in the interior.

evidently Hervé remained based at Saint Cloud and things seem to have calmed down for a while until June 1950, by which time relationships had deteriorated to such a point that Louis wrote to Hervé:

après de longs mois de réflexion je suis arrivé à la conclusion que j'avais commis une faute en voulant avoir mes deux fils dans mes affaires commerciales. J'avais cru que cette association après les longues années de séparation nous aurait rapprochés.... Je considère qu'il

Chapter Eight

est nécessaire que vous vous trouviez une situation en dehors de mon contrôle (After long months of reflection I have concluded that I made a mistake in wishing to have my two sons in business with me. I believed that after long years of separation it would have drawn us together.... I consider that it is necessary that you should find yourself a job outside my control).

He emphasised the point by stressing how much better his relationship with Jean had become since the latter had moved on, although it seems that the Jean may not have shared this view.

Hervé replied, expressing surprise, saying he was under the impression that their personal relationship had much improved since September and that business was looking up, but he would leave at the end of the month if that was what his father really wanted. He did not leave and by May 1951 matters had been patched up again and Hervé was confirmed in his position as Managing Director of la Societé K.L.G. and entitled to a 5 per cent share of the profits. Business in the early 1950s was reasonably successful, but between 1953 and 1959 the volume of sparking plugs sold diminished by a third. This was largely a reflection of the fact that K.L.G. prices in France were higher than

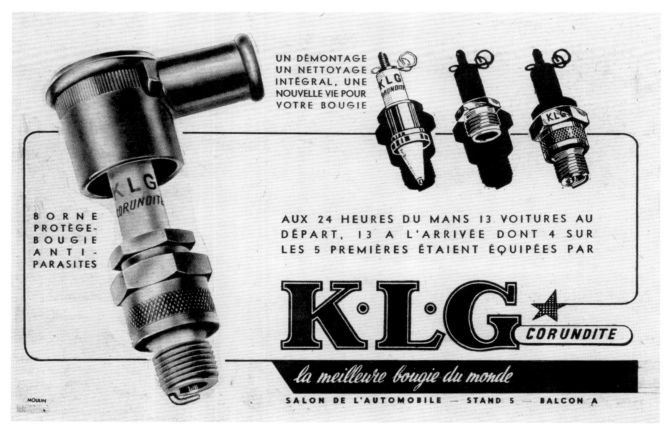

1951 advertisement for K.L.G. plugs showing how they could be dismantled for cleaning.

Hervé Coatalen greets a customer on the K.L.G./Lockheed stand at the 1951 Salon de l'Aéronautique, Paris.

their competitors, which meant they were not specified as original equipment by any French car manufacturers.[7]

To add to their problems, in 1960 a broker by the name of Lambert, who had acquired shares in K.L.G. cheaply, was asking awkward questions at the Extraordinary General Meeting as profits continued to decline. It is not surprising that, soon after Louis Coatalen's death in 1962, his son Hervé sold off the French K.L.G. business and took early retirement himself.

Louis Coatalen never retired fully but remained active until the end. The Société Française des Freins Hydrauliques Lockheed (SFFHL) business grew from one factory at Saint-Ouen until it required two further factories. In 1957, when Coatalen was

K.L.G. Société Anonyme letter signed by Louis Coatalen, 6 June 1957.

---

7. Information from Coatalen family documents.

Chapter Eight

Top left: K.L.G. promotional cufflink and tie pin made with miniature spark plugs.
Top right: K.L.G. spark plugs were supplied in distinctive yellow tins.
Bottom left: Another K.L.G. promotional device was this especially designed pack of cards.
Bottom right: Jean Coatalen designed and patented this moulded plastic packaging for individual spark plugs in 1955.

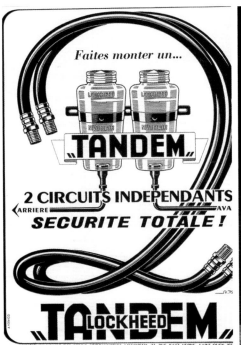

Left: Advertisement for Lockheed's twin-circuit hydraulic braking system.
Right: Louis Coatalen in old age still taking part in an official line-up to greet important visitors (event and people not identified).

seventy-eight years old, SFFHL purchased a factory site at the Zone Industrielle du Pont d'Arcole, Beauvais, which initially employed 300 people. Two years later there were 1,000 employees on this site. The town named the street leading to the factory 'rue Coatalen'. To judge by the number of patents he took out between the end of the war and his death, this was one of the most productive periods of his life. More than sixty patents were granted in his name over the seventeen-year period. The vast majority were connected to improvements in hydraulic braking systems and included a number relating to disc brakes. They went from concepts such as a hydraulically driven pump or compressor, down to the details of master cylinders, pistons, piping systems and even brake fluid based on castor oil mixed with polypropylene glycol. However, in October 1961, when he was over eighty years old, Coatalen finally sold his shares in SFFHL to the DBA group (Ducellier, Bendix, Air Equipment), which, in turn, some thirty-five years later, was absorbed by Bosch. At the time Coatalen sold out, the business was employing 2,300 people. He retained the title Directeur Technique of the Lockheed division until his death.[8]

When he was in his seventies, Coatalen was regarded as one of the 'grand old men' of the automobile industry. In 1953 he was elected President of La Société des Ingénieurs de l'Automobile (SIA), a post he held for three years, in succession to his friend Henri Perrot. He had been a member of the SIA, which

8. In 2014 the factory buildings were destroyed and the rue Coatalen led nowhere.

Chapter Eight

1951 Lockheed hydraulic brake advertisement with tandem master cylinders giving two independent braking circuits.

Perrot had helped to found in 1927, since 1934. The Spanish government recognised his contribution to their own parallel organisation, the STA, by giving him the Civil Order of Merit and in 1954 he was promoted to the rank of Officier de la Légion d'Honneur, which was recognition by the French government and wider society of his long career in the industry. The following year he was made an honorary member of the Italian Associazione Tecnica Automobile. He also remained the SIA representative at the Fédération Internationale des Sociétés d'Ingénieurs des Techniques de l'Automobile. When he was eighty years old, an article appeared in the French motoring press reviewing his career under a headline that stressed that he was still at the helm.

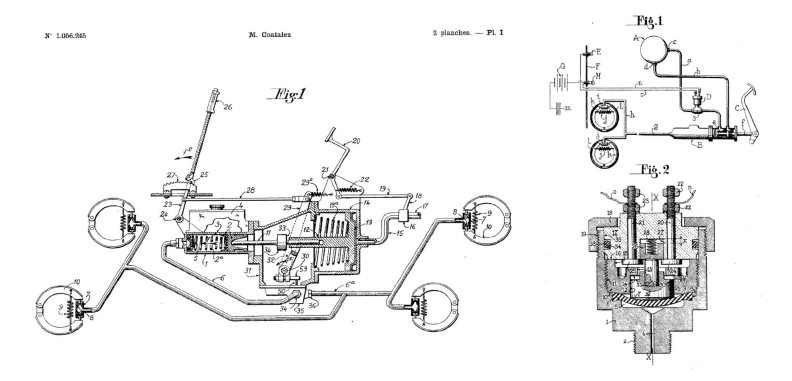

Top left: Diagram from 1954 patent no. FR1056245 for a perfected braking circuit installation.
Top right: Diagram from 1956 patent FR1125022 for a pressure failure warning light system.
Bottom: A meeting of some of the 'grand old men' of the automobile industry at the smart Parisian restaurant La Tour d'Argent c. 1955. Louis Coatalen (1879–1962) is front left with clockwise Jean Coatalen (1916–76), Jean Panhard (1913–2014), Hervé Coatalen (1913–99), Henri Perrot (1883–1961), Charles Faroux (1872–1957) and Paul Panhard (1881–1969). Originally built as an open terrace in the 1930s with a spectacular view of the Seine, this part of the restaurant had by then been glazed but the unusual cantilever chairs had been retained.

Chapter Eight

Private Life

As he grew older Coatalen allowed himself to relax a little, although he never stopped working. In the 1950s Coatalen bought himself another yacht *Trylona*, designed by the same naval architect as *Tirrena* that he had commissioned prior to the war, as there had been a compensation payment but which had to be spent in Italy.

Carole described their life on board *Trylona* during the summer in the Mediterranean:

> Mostly the daily programme was as follows. We would get up around eight o'clock. Then my grandfather would have a long conversation with the captain about the weather forecast. The captain was not always keen to take the boat out and would say, 'oh, monsieur, it is windy you know, we shouldn't go'. But as soon as my grandfather saw the sun shining he would insist and we would leave the port and drop anchor off the beach at Paloma or in front of Menton. At the end of the morning a thermometer was dipped in the water to check the temperature before we climbed down the ladder to bathe. My grandmother and I loved the contact with the water but my grandfather did not always swim.

His wife Dickie was an elegant companion to Louis Coatalen in his later years. Carole remembers her as always very smart and wearing a 'long interior dress' and smoking through a cigarette holder with her box of cigarettes and ashtray constantly by her side. She went once a week to the hairdresser to have her blue-grey rinse maintained. She evidently enjoyed life to the full after World War II, as she loved being on their yacht in the Mediterranean sailing out of Cannes or Monaco or driving fast in her black Simca sports car. She played the piano, was an accomplished horsewoman and fulfilled the role of lady of leisure, taking breakfast in bed before dressing in clothes laid out by the maid and taking tea in the afternoon with one or two guests.[1]

---

1. Carole de Chabot, *Rebelle avec l'aide de Dieu*, Memorizon, Champigny sur Marne, 2012, p. 49.

Louis Coatalen was at the Lockheed works when he had a brain haemorrhage and three days later he died on 23 May 1962, aged eighty-two. A private funeral was held on 25 May, followed by a memorial service on 4 June at the church of Saint-Honoré d'Eylau attended by many friends. He is buried in the graveyard at Jouars near Pontchartrain. His wife, Ellen Mary 'Dickie' survived him by nearly eleven years and was buried by his side in 1973.

An obituary in *The Times* recalled briefly his career and the racing successes of Sunbeam and Talbot-Darracq cars but noted that 'in spite of his undoubted ability as a designer, Coatalen appears to have been rendered impatient by the success of certain of his rivals, and was somewhat prone in consequence to copy their methods rather than adhere to his own'. A few days later, Lord Semphill, who had been apprenticed to Rolls-Royce in 1910 and subsequently had become a pilot and was later President of the Royal Aeronautical Society, wrote to point out that Coatalen's death removed 'one of the three leading designers responsible for the engines that powered the aircraft and airships used by the Royal Naval Air Service in the First War', and that the RNAS remembered him with gratitude as a designer ahead of his time. On the same day *The Times* also published a letter from the Sunbeam racing historian, Anthony S. Heal, who wrote of Louis Coatalen, 'If, as your correspondent suggests, his Gallic impatience led him on occasion to copy the work of some of his rivals, this practice certainly proved successful in some of the most important international motor races.... But apart from his claim to fame as a designer, Coatalen's great contribution to the Sunbeam Motor

Louis Coatalen's last yacht *Trylona* in San Remo harbour in 1957.

Chapter Eight

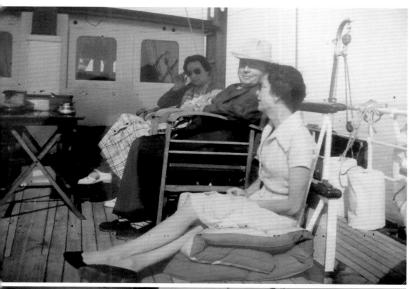

Top: Coatalen aboard *Trylona* with Dickie and his daughter-in-law Anna in 1957.
Bottom: Louis with his son Hervé on board *Trylona* on the same occasion.

Company was surely as an impresario. He led and inspired others to achieve miracles they would not themselves have believed possible.'[9]

This reputation for borrowing designs from elsewhere and seeking to improve them seems to have stuck with Coatalen more than it did to some of his contemporaries who have perhaps had more loyal defenders. It is interesting that Donald Bastow, in his book *W.O. Bentley – Engineer*, wrote the following, which could equally well have been written about Louis Coatalen:

> The professional engineer knows that there is no merit in originality as such, and anyone designing any kind of mechanism can be helped a great deal by study of what has gone before ... accepting and improving the good and rejecting the bad ... to insist on complete novelty would be a mark of vanity from which W.O. certainly did not suffer.[10]

John Wyer (an ex-Sunbeam apprentice and successful motor-racing manager) certainly felt that Coatalen got unduly criticised. He was moved to write that the story of the copying of the Peugeot in 1913, published in *Motorsport* in 1977, was

> not improved by Boddy's peevish asides about Coatalen. Boddy would be a better journalist if he did not allow his opinions and prejudices to show through quite so much.... I have never been able to understand why the purists get so neurotic about this

---

9. *The Times*, 25 May 1962 and 1 June 1962.
10. Donald Bastow, *W.O. Bentley – Engineer*, Haynes, 1978, p. 28.

> Peugeot business. If your rival has a better engine or car than you do, it seems to me to be a perfectly natural and sensible thing to copy it and try to improve on it. It is really not necessary to invent the wheel every time you produce a new car and if nobody ever copied anyone else there would have been a lot less progress.

He went on to say that although 'nobody would maintain that Coatalen was a great engineer ... he was an entrepreneur at a time when the technology was in its infancy and the entrepreneurs were in full bloom'. Why did only Sunbeam get criticised? Wyer pointed out that Jano had left Fiat to join Alfa Romeo and then 'for the next 20 years he continued to design engines which were clearly derived from his Fiat experience. Yet I have read no criticism of Alfa Romeo.'[11]

In 1974 the Paris Mint (Monnaie de Paris), at the suggestion of the Commission Historique de l'Automobile Club de France, issued a series of medals to commemorate the pioneers of the automobile industry. Included among the famous names Armand Peugeot, Louis Renault, André Citroen and Louis Delage was also Louis Coatalen. His portrait was sculpted by Claude Lhoste and the obverse of the medal illustrates the fact that in 1923 his engines conquered not only on land but also in the air and on water.

Coatalen's reputation was severely blighted (pardon the pun) when Anthony Blight published his great work *Georges Roesch and the Invincible Talbot* in 1970. Roesch who, perhaps understandably, had a chip on his shoulder that his own contribution had been undervalued for too long, cast Coatalen as the villain responsible for frustrating his ambition.[12] Blight reflected and embroidered this viewpoint with his colourful prose. For example, he wrote of the formation of the STD group: 'If the Talbot Darracq marriage was a spiritual matter, the formation of STD had an almost daemonic basis. It can be traced back to a single source: the burning ambition of Louis Coatalen to build a winning Grand Prix racing car.' Since then more information about the formation of the group has emerged and we now know that Darracq acquired Talbot in order to repair army surplus vehicles (hardly a spiritual marriage) and the merger with Sunbeam was necessitated following the death of the chairman John Marston.

Concerning events in the mid-1920s, Blight also wrote,

> the strong arm which had already lost its grip on Sunbeam was beginning to weaken in a general sclerosis. In fifteen years of frenzied activity, Coatalen had squandered his physical resources as recklessly as the material resources of STD. By the Autumn of 1925 both had been drastically curtailed; the boundless energy and enthusiasm which had

---

11. John Wyer, letter to Anthony Heal, 1 December 1977 (author's collection).
12. Maurice Platt, who knew Georges Roesch well, wrote of him that, 'Like all the great creative artists in engineering ... he was a supreme egotist, incapable of making any concessions that threatened the integrity of his concepts.' *An Addiction to Automobiles*, F. Warne, 1980, p. 41.

Chapter Eight

brought this great engineer international fame and rewards had burned itself out, and he had embarked on a melancholy journey of personal decline which was to bring him to the verge of insanity and terminate his association with England and STD for ever.[13]

It now seems clear that this assessment was exaggerated and premature. The curtailment of the racing teams and the apparent sclerosis in terms of development was mainly a reflection of the financial situation of the group which was not of Coatalen's making but stemmed directly from the structure of the group of companies set up in 1920.

More positively, that great New Zealand connoisseur of Sunbeam cars and expert in leadership, David Adams, summed up Coatalen as a man with 'vision and true leadership qualities. He was able to share this vision with others who then bought into it. He was able to delegate yet retain control sufficiently well to get remarkable outcomes. He must have understood well the strengths (and probably the weaknesses) in his people to get the best from them. He knew where he wanted to go and was able to select and lead people to enable him to get there.'[14]

Undoubtedly, Coatalen became disillusioned by the way the STD Group had evolved and frustrated at no longer being able to pursue the policies that had previously brought him success. His contribution was not helped by his 'lifestyle choices', which must have reduced his impact on the business

The commemorative bronze medal featuring Louis Coatalen and his engines issued in 1974.

13. Anthony Blight, *Georges Roesch...*, p. 22.
14. David Adams, email to the author, 4 August 2009.

as time went on. It is evident that by the end of the 1920s his thinking appears confused but at this distance in time it is not possible to know what influences were then at work or even to ascertain the exact nature of his illness. The fact that, at the very same time that his life and career were in turmoil he had the presence of mind to establish the basis of his future work, suggests that he had not lost his sanity but had developed a clear idea of where he wanted to be and how he wanted to spend the final part of his career. Over the years that followed he rebounded and built up another successful business at Lockheed in a manner that mirrored his earlier successes at Humber and Sunbeam. A man who constantly sought better ways of engineering and to whom more than 150 patents were granted over his working life cannot simply be written off as a plagiarist! Even if he was an idol with evident feet of clay, he deserves to be remembered as 'an engineering impresario who led and inspired others to achieve miracles they would not themselves have believed possible'.

Appendices

Appendix One
Louis Coatalen – Personal Competition Achievements

Results are given as stated in event reports of the time.

Date	Event	Car	Bore & Stroke (c.c.)	Place
1905				
29.04.05	London to Coventry	8-10 Humber	3⅛" x 3¾", (1885)	
10.06.05	South Harting Hill-Climb	8-10 Humber		1st
	Brighton Speed Trials:			
20.07.05	Scratch Race for cars £200-£350	8-10 Humber		1st
21.07.05	Handicap Sweepstake for standard cars	"		
22.07.05	Handicap Sweepstake for touring cars	"		
05.08.05	*Blackpool Race Meeting:*			
	Lancashire Handicap	8-10 Humber		1st
August?	Birdlip Hill Wager	8-10 Humber		
August	Shelsley Walsh Hill Climb	8-10 Humber		4th on handicap
September	*Skegness Sands:*			
	Mile Handicap Race – 3rd heat	8-10 Humber		1st
	Mile Scratch Race – cars under £350	"		1st
	Mile Scratch Race – cars any HP	"		
September?	Snake Hill-Climb, Glossop, Derbys	8-10 Humber		3rd
1906				
24.05.06	Herefordshire AC Frome's Hill	10hp Humber	3½" x 3¾"	
23.06.06	South Harting Hill-Climb	10-12 Humber		2nd
27.09.06	Isle of Man TT Race	20hp Humber	102 x 102 mm	6th
12.10.06	Blackpool Race Meeting	16-20 Humber	100 x 100, (3142)	2nd
1907				
30.05.07	Isle of Man TT Race	20hp Hillman-Coatalen	108 x 114	retd
31.05.07	Graphic Trophy Hill-Climb	20hp Hillman-Coatalen		7th
June	Henry Edmunds Trophy	Hillman-Coatalen		8th
1908				
24.09.08	Isle of Man TT Race	4in Hillman-Coatalen	4"x 5", 102 x 127mm	9th
	"	"		retd

Louis Coatalen – Personal Competition Achievements

Driver	Distance	Time	Speed	Remarks
Louis Coatalen	90 miles	3h 25m		Only top gear
LC	1833 yards			Mr JW Adams' car, won Class A cars £150-£350, 4 passengers
LC	1 mile			Mr JW Adams car
LC				Beaten by FD Lyon's Panhard
LC				Beat Astell's 15hp Orleans in first heat but eliminated after
LC	1.5 miles	2m 16.4s		Won all 3 heats. Won by 10.6s on handicap. Mr J Newton's car
LC	1899 yards	4m 20s		With 3 passengers, same car as Brighton
LC	1133 yards	2m 20.6s		Mr Adams' car
LC	1 mile			Mr Adams' car, won by 17 yds
LC	1 mile			Alldays & Onions 2nd. Won by 16.2 secs, 10 entries
LC	1 mile			Walk over in first heat, beaten in semi final
LC	3 miles			Class 2
LC	1289 yards			"Very rapid ascent indeed"
LC				Class A cars between £150-£350. beaten by Stanley steam car
LC		5h 0m 52.8s	32.1 mph	Dirt in carburetter
LC	1 mile			Standing mile, beaten by 40hp Bianhi.
LC	240 miles			32cwt 75lbs, broken rear spring after accident, retired end of 2nd lap, set up fastest lap.
LC	2550 yards	4m 37.8s		Slieau Lewaigue Hill
LC		3m 51.2s		1st run 1m 54.4s. 2nd run 1m 56.8s. Carter's Hill, Under river.
LC	339.5 miles	8h 20m 35s	40mph	
KL Guinness				Lap 4, broken frame

Appendix One

Date	Event	Car	Bore & Stroke (c.c.)	Place
1909				
June	Scottish Trial 1000m 6 day	14-20 Sunbeam		
17.07.09	Shelsley Walsh Hill Climb	14-20 Sunbeam		
02.10.09	*Wolverhampton AC Hill Climb:*			
	Harley Bank	Sunbeam 16-20		1st in class
	"	40hp Hillman-Coatalen		2nd
1910				
28.03.10	*Brooklands Easter Meeting:*			
	Raglan Cup Race	12-16 Sunbeam, 4 cyl	80 x 120, (2412)	1st
	March Handicap	Nautilus 21hp, 4cyl	92 x 160, (4257)	
27.04.10	Brooklands Raglan Cup Race	12-16 Sunbeam	80 x 120, (2412)	2nd
	"	"	"	3rd
11.06.10	Welbeck Flying Kilometre	16-20 Nautilus	92 x 160, (4257)	
18.06.10	*Brooklands June meeting:*			
	4th 21hp Rating Handicap	Nautilus' 21hp, 4cyl	92 x 160, (4257)	
	Invitation Race	12-16 Sunbeam	80 x 120, (2412)	2nd
	"	"	"	4th
	First 16hp Rating Race	12-16 Sunbeam	80 x 120, (2412)	1st
	"	"	"	2nd
25.06.10	*Yorkshire AC Speed Trials:*			
	Saltburn Sands	12-16 Sunbeam, 4 cyl	80 x 120, (2412)	2nd Class 1
02.07.10	*Shelsley Walsh Hill-Climb:*			
	Henry Edmonds Trophy	12-16 Sunbeam, 4cyl	80 x 120, (2412)	3rd
10.09.10	*Wolverhampton AC:*			
	Ironbridge Hill-Climb	12-16 Sunbeam, 4cyl	80 x 120, (2412)	1st Class 2
05.10.10	*Brooklands October Meeting:*			
	2nd 16hp Rating Handicap	12-16 Sunbeam, 4cyl	80 x 120, (2412)	1st
	76mph Handicap	"	"	2nd
	Over 60mph Cars Race	"	"	3rd
06.10.10	*Brooklands Class Record:*			
	Flying half-mile	12-16 Sunbeam	80 x 120, (2412)	

Louis Coatalen – Personal Competition Achievements

Driver	Distance	Time	Speed	Remarks
LC	1000 miles			Class F (£425-£525) 2nd
LC		3m 4s		Stopped to release needle on carb
LC				One of 4 running, Bayliss, Genna, Cureton, Coatalen
LC		2m 15s		40hp H-C 2 str, 6cyl, 5" bore, beaten by Lisle 15hp Star. A 40hp H-C time of 2m 30s also recorded H. Nelson Smith.
LC	8.5 miles		55.5 mph	First appearance of 12-16 Sunbeam at Brooklands
LC	11 miles			Baulked by accident, lapped at 75.5mph
LC / NF Bayliss				
LC	1 km		82mph	Duke of Portland estate, Clipstone, Notts AC.
LC	8.5 miles			Retired, engine overheated
LC / NFB	5.75 miles			
LC / NFB	5.75 miles		61.25 mph	Won by 5 yds from Bayliss
LC				LC won two medals. NF Bayliss won 2 gold medals.
LC		1m 34.4s		
LC	c. 0.75 mile			RAC silver medal for best performance irrespective of class, Star & Sunbeam Trophies, 2 gold medals on formula.
LC	5.75 miles		63mph	From scratch, 3 starters
LC	8.5 miles			14s behind winner, 12 starters
LC				
LC	0.5 mile f/s	25.016s	71.982mph	New record 16hp rating

Appendix One

Date	Event	Car	Bore & Stroke (c.c.)	Place
1911				
25.03.11	*Brooklands Match Race:*			
	Sunbeam v. Singer	Toodles II	80 x 160, (3217)	
17.04.11	*Brooklands Easter Meeting:*			
	Sunbeam v. Napier Match Race	Toodles II	80 x 160, (3217)	1st
	4th 76mph Handicap	"	"	3rd
	4th 100mph Handicap	"	"	2nd
	Easter Sprint Race	"	"	2nd
18.04.11	*Brooklands Class Record:*			
	16 Rating Short Record	Toodles II	80 x 160, (3217)	
10.05.11	*Brooklands May Meeting:*			
	5th 100mph Handicap	Toodles II	80 x 160, (3217)	3rd
	May Sprint Race over 70mph cars	"	"	1st
	Winners Handicap	"	"	1st
22.07.11	*Wolverhampton & District AC:*			
	Ironbridge Hill Climb	Sunbeam 12-16		1st
29.07.11	Brooklands Inter Club Meeting	15.9 Sunbeam		
01.09.11	Brooklands 12 Hour Record	Toodles IV 25-30 Sunbeam 6cyl, 30.1 hp	90 x 165? (6107)	

Louis Coatalen – Personal Competition Achievements

Driver	Distance	Time	Speed	Remarks
				Non started, broken crankshaft
LC	5 miles		83mph	Beat 60hp Napier by 15.4 secs
LC	8.5 miles			1st Boillot, Lion Peugeot
LC	8.5 miles			1st Kane, Imperia by 15.9s
LC	2 miles			1st Boillot, Lion Peugeot
LC	0.5 mile f/s	20.892s	86.157mph	New 16hp record
LC	8.5 miles			Protest dismissed
LC	2 miles		74mph	
LC	5.75 miles		79mph	Won by 8 lengths
LC	0.75 miles			Won both open and closed events
LC		14.2s		L.C. was part of Yorks AC team, won relay
LC & TH Richards	400 miles	5h 20m 30s	74.87 mph	
	500 miles	6h 40m 16s	74.95 mph	
	600 miles	7h 57m 59s	75.32 mph	
	700 miles	9h 16m 34s	75.46 mph	
	800 miles	10h 34m 29s	75.65 mph	
	900 miles	11h 53m 36s	75.68 mph	
	miles, yds			engine ran non-stop
	300m1421y	4h	75.2 mph	
	373m135y	5h	74.21 mph	
	451m445y	6h	75.2 mph	
	525m566y	7h	75.05 mph	
	602m975y	8h	75.32 mph	
	678m138y	9h	75.34 mph	
	757m28y	10h	75.71 mph	
	832m704y	11h	75.72 mph	
	907m1535y	12h	75.66 mph	Fastest lap (103rd) 79mph

Appendix One

Date	Event	Car	Bore & Stroke (c.c.)	Place
04.10.11	*Brooklands October Meeting:*			
	2nd 100mph Long Handicap	Toodles IV, 6cyl 30.1hp, 25-30	90 x 160, (6107)	1st
	2nd 100mph Short Handicap	Toodles IV	90 x 160, (6107)	4th
	15.9hp rating Standard Car Race	15.9 Sunbeam	80 x 149, (2996)	1st
	"	"		2nd
	October Sprint Race	Toodles IV	90 x 160, (6107)	2nd
07.10.11	*Brooklands Records:*			
	16hp Rating Class Long Record	Toodles II, 15.9hp 4cyl	80 x 160, (3217)	
	40hp Rating Class Long Record	Toodles IV, 25-30, 6cyl	90 x 160, (6107)	
1912				
05.08.12	*Brooklands Bank Holiday Meeting:*			
	8th 100mph Short Handicap	Toodles IV, 30.1hp	90 x 160, (6107)	1st
10.08.12	*Yorkshire AC:*			
	Pateley Bridge Hill-Climb	Coupe de l'Auto	80 x 149, (2996)	1st equal
31.08.12	*Woodhouse Eaves, Leics AC:*			
	Beacon Hill-Climb	Coupe de l'Auto	80 x 149, (2996)	2nd
18.09.12	Brooklands 4 & 5 Hour Record	Coupe de l'Auto Sunbeam, 4cyl	80 x 149, (2996)	
1913				
06.04.13	*Monaco Motor Boat Meeting:*			
	Prix du Premier Pas	Fuji Yama III	80 x 120, (2413)	1st
03.05.13	*Lancashire AC:*			
	Rivington Pike Hill-Climb	1913 GP Sunbeam 6cyl	80 x 150, (4524)	1st
13.05.13	Brooklands Class C records	Short stroke C de l'Auto	80 x 120, (2413)	
07.06.13	Shelsley Walsh Hill-Climb	V8 Sunbeam	80 x 150, (6032)	6th
1914				
	Monaco Motor Boat Meeting:			
09.04.14	First 21ft class race	Toto 4 cyl 16 valve, tohc	81.5 x 117, (2440)	1st
10.04.14	25 km race	"		1st
11.04.14	25 km race	"		2nd
15.04.14	Omnium handicap	"		4th
18.04.14	Coupe Hirondelle	"		retd

Louis Coatalen – Personal Competition Achievements

Driver	Distance	Time	Speed	Remarks
LC	8.5 miles		81.75 mph	
LC	5.75 miles			
LC	50 miles		58.25 mph	
NFB				
LC	2 miles			
LC	10 laps	20m 56.34s	79.29 mph	
LC	10 laps	20m 30.65s	80.90 mph	
LC	5.75 miles		82.5mph	
LC		ftd (equal)		Equal w. J. Higginson's 80 hp La Buire
LC	1453 yards	65.6s		Class A
LC, RF Crossman & D Resta	319m 242yds	4h	79.78mph	World Record
	391m 1429yds	5h	78.36mph	World Record
	400 miles	5h 5m 53s	78.46mph	World Record
LC				Ernest Martin's boat
LC				On formula, silver cup, unlimited class
LC	10 laps	20m 6.57s	82.55mph	Unofficial records running on benzole.
LC		65.0s/60.2s		V8 aero engine in standard chassis
LC		56m		J.A. Holder's boat, LC with J. Chassagne lead from start
LC	25 km			Fastest lap: 31.4 knots. Handicap: 1m 48s
LC	25 km	25m 48s	31.5 knots	Handicap: 3m 36s
LC				
LC				Petrol tank ripped off by wave

Appendix Two
Louis Coatalen Patents (Great Britain and France)

Many patents were also taken out in other countries but as these replicate the original patents they have been omitted from this list.

Publication Date	Patent No.	Title	Louis Coatalen with	Notes
23.12.1912	GB191206490	Improvements in starting systems for internal combustion engines	SMCC	Compressed air starter, no dead centre, multicylinder 120° cranks, reduction gear drive
28.10.1915	GB191500078	Improvements in multicylinder fluid pressure engines	SMCC	Double V engine on common crank case. Strong, light, small multicylinder engine
12.08.1915	GB191500077	Improvements in driving mechanism for aerial propellers	SMCC	Variable speed gearing for props
29.07.1915	GB191500117	Improvements in ambulance waggons and the like	SMCC	Foldaway stretcher supports or bench seat
28.10.1915	GB191501931	Improvements in multicylinder fluid pressure engines	SMCC	A development of 191500078 providing for 4 V-engines (narrow or wide V) on common crankcase with 4 crankshafts driving central drive shaft
02.12.1915	GB191503324	Improvements in valve-operating mechanisms for internal combustion engines	SMCC & HCM Stevens	Double groove switchover cam to operate exhaust and inlet valve from single cam
02.03.1916	GB191609666	Improvements in cylinders for internal combustion engines	SMCC	Light welded cylinder construction with side valve
02.03.1916	GB191510012	Improvements in lubrication systems for internal combustion engines	SMCC	Oil pump in upper crankcase
04.05.1916	GB191510011	Improvements in valve and similar operating mechanism for internal combustion engines	SMCC & HCM Stevens	Cam to operate on alternate revolutions
10.08.1916	GB191514578	Improvements in aeroplane driving mechanism	SMCC	Gears for propellers in multi-engine planes
18.01.1917	GB103165	Improvements in lubrication of cam-shafts for internal combustion engines	SMCC	Lubrication hole through back of cam
18.01.1917	GB103164	Improvements in pistons for internal combustion engines	SMCC	Oil collection holes through piston skirt
18.01.1917	GB103163	Improvements in cam-shaft-driving mechanism for internal combustion engines	SMCC	Wide spur gear to drive narrow gears to save space on overhead camshaft drive

Louis Coatalen Patents

Publication Date	Patent No.	Title	Louis Coatalen with	Notes
18.01.1917	GB103162	Improvements in lubrication systems for internal combustion engines	SMCC	Double oil pump to provide high and low pressure feeds
18.01.1917	GB103161	Improvement in joints for link-work	SMCC	Lubrication and connection of articulated connecting rods
18.01.1917	GB103160	Improvements in camshaft casings for internal combustion engines	SMCC	Improved access to overhead camshaft for rocker removal and to stop lubricant flowing down tappet guides
18.01.1917	GB103159	Improvements in the construction of crank chambers for aviation engines	SMCC	Reinforcing webs cast into crankcase for propeller spindle case
18.01.1917	GB103158	Improvements in starting mechanism for internal combustion engines	SMCC	Starting gears for more than one engine
19.03.1917	GB104897	Improvements in internal combustion engines for aeroplanes	SMCC	Shaped shield for V engines between block and exhaust system
19.03.1917	GB104896	Improvements in internal combustion engines	SMCC	Tohc engine with aluminium cylinder block with dry liners and detachable aluminium head with bronze valve seats. Central sparking plugs. Removable finger cam followers.
19.03.1917	GB104895	Improvements in multi-cylinder internal combustion engines	SMCC	Inlet manifolding for W engines
19.03.1917	GB10894	Improvements in V-type internal combustion engines	SMCC	Easy removal of pumps and dynamo by grouping
10.04.1919	GB124808	Improvements in fuel feed systems	SMCC	Vane wheel driven vacuum fuel pump
24.04.1919	GB125425	Improvements in internal combustion engines	SMCC	Narrow V engine cast as one block with common water jacket for strength and lightness
03.09.1917	GB109285	Improvements in valve operating mechanism for internal combustion engines	SMCC	
05.10.1917	GB110022	Improvements in oil filters on internal combustion engines	SMCC	Individual removable filters between main oil supply and bearing
23.10.1917	GB110575	Improvements in internal combustion engines	SMCC	Sump casing to act as funnel to collect air and provide warmed air to carburettor

Appendix Two

Publication Date	Patent No.	Title	Louis Coatalen with	Notes
15.05.1919	GB126390	Improvements in cylinders for internal combustion engines	SMCC	Light aluminium cylinder block with shrunk in dry liners and screwed in valve seats
15.05.1919	GB126389	Improvements in valve operating mechanism for internal combustion engines	SMCC & HCM Stevens	3 valve sohc rocker assembly
12.06.1919	GB127649	Improvements relating to internal combustion engines	SMCC & HCM Stevens	Countershafts and bevel wheels to drive camshafts and ancillaries on narrow V engines
12.06.1919	GB127672	Improvements relating to internal combustion engines	SMCC	Valves radiating from centre opened by bell crank levers
27.09.1920	GB151715	Improvements in heaters and baffles for the induction pipes of internal combustion engines	SMCC & J.S. Irving	Combined air heater and baffle for induction
27.09.1920	GB151714	Improvements in luggage grids for use on motor vehicles	SMCC & J.S. Baugh	Means for locking in place folding grid
27.09.1920	GB151713	Improvements in clutches for power transmission	SMCC & HCM Stevens	Positive drive
27.09.1920	GB151377	Improvements in crankshafts for fluid pressure engines	SMCC & HCM Stevens	Built-up crankshaft coupling
10.02.1921	GB158425	Improvements in movable seats for use on vehicles	SMCC & JA Cooper	Tip-up adjustable seats
22.02.1921	GB159301	Improvements in folding seats for use on vehicles	SMCC & JA Cooper	Folding seat
03.03.1921	GB159402	Improvements in packing for stuffing boxes and the like	SMCC & JS Irving	Graphite impregnated cork
06.12.1921	GB172409	Improvements in or relating to cam shafts of internal combustion engines	SMCC & JS Irving	Regular torsional resistance and silencing through oil pump
24.11.1921	GB171534	Improvements in transmission gearing for use on aircraft, motor boats and the like	SMCC	Gearing in transmission with multiple engines
26.10.1922	GB187465	Improvements in removable windows and fittings thereof for use on motor vehicles	SMCC & CB Kay	Easily fitted side-screens
12.10.1922	GB186850	Improvements relating to detachable windows for use on motor vehicles	SMCC & CB Kay	Providing access to internal door handles

Publication Date	Patent No.	Title	Louis Coatalen with	Notes
03.08.1923	GB201597	Improvements in antifriction bearings	HCM Stevens (named first) LHC & SMCC	Roller race casing design
01.01.1924	GB208754	Improvements in friction transmission clutches	SMCC & HCM Stevens	Improved clutch
20.12.1923	GB208308	Improvements in vehicle bodies	SMCC & JA Cooper	Slots for improved sealing of side curtains
20.09.1923	GB203940	Improvements in vehicle bodies	SMCC & JA Cooper	Improved folding hoods
10.01.1924	GB209250	Improvements in brakes for use on vehicles	SMCC & HCM Stevens	With servo
28.02.1924	GB211629	Improvements in brakes for use on motor vehicles	SMCC	Main shoe and servo shoe
03.01.1924	GB208919	Improvements in or relating to electric switches actuated by the movement of a vehicle door	SMCC & H.M. Weir	Interior light activated by door
16.04.1925	GB232021	Improvements in valve operating mechanism for internal combustion engines	SMCC	Eccentric rocker bushes for adjustment
18.06.1925	GB235359	Improvements in internal combustion engines	SMCC & Alfred Huggins	Supercharger jacketed by fuel supply to cool blower and warm fuel
18.06.1925	GB235358	Improvements in plate clutches	SMCC & HCM Stevens	Clutch with 'very sweet action'
13.08.1925	GB238026	Improvements in metal casting	SMCC & JS Irving	Steel tubes cast into aluminium allowing for different rates of expansion
28.05.1925	GB234307	Improvements in water cooling systems for internal combustion engines	SMCC & HCM Stevens	Thermostatic flow control
17.09.1925	GB239669	Improvements relating to fuel supply systems for internal combustion engines	SMCC	Blower control

Appendix Two

Publication Date	Patent No.	Title	Louis Coatalen with	Notes
20.08.1925	GB238363	Improvements relating to fuel supply systems for internal combustion engines	SMCC	Blower activated at full throttle
03.12.1925	GB243560	Improvements in carburettors for internal combustion engines	SMCC & H Wilding	Baffle to improve atomisation
06.07.1926	GB254408	Improvements relating to mechanism for varying the angular relationship of two shafts relative to a third	SMCC & SH Attwood	Fuel injection timing and valve timing adjustment through epicyclic gears
06.07.1926	GB254788	Improvements relating to valve operating mechanisms	SMCC & HCM Stevens	Angular relationship of cams to be varied
23.12.1926	GB262973	Improvements in the bearings of leaf springs for motor vehicles	SMCC & JS Irving	Perforated floating bush for lubrication
04.10.1928	GB297948	Improvements in means for supplying liquid fuel to internal combustion engines	SMCC & SH Attwood	Timed delivery of oil, but variable supply under constant pressure
23.04.1929	GB310391	Improvements relating to vehicle springing	SMCC	Independent Front Suspension design
10.04.1930	GB327565	Improvements relating to fuel injection devices for internal combustion engines	SMCC & SH Attwood	Fuel pump suction valve actuation
17.07.1930	GB332043	Improvements relating to internal combustion engines of the compression	SMCC & SH Attwood	Improved mounting of fuel injection valve
10.04.1930	GB327646	Improvements in means for controlling internal combustion engines of compression ignition type	SMCC & SH Attwood	Fuel injection valves actuated through levers cooperate with air inlet and exhaust cams
25.09.1930	GB335350	Door locks	SMCC	Allows full floating engagement of bolt with striker plate without slamming
09.07.1931	GB352333	Pistons	SMCC	One piece ferrous skirt with gudgeon pin bosses pressed from skirt

Louis Coatalen Patents

All previous patents were taken out jointly with the Sunbeam Motor Car Co. and sometimes with another named person. All patents listed from here on were taken out by Louis Coatalen alone. After 1931 patents were taken out first in France.

Publication Date	Patent No.	Title	Louis Coatalen address	Notes
17.07.1930	GB332055	Means for starting internal combustion engines and driving electric generators	Rue Spontini	Compressed air starter, quiet and without shock
10.07.1930	GB331700	Hydraulic brakes	France	Dual master cylinders
14.08.1930	GB333365	Hydraulic brakes	France	Means for preventing oil loss in event of leak
23.10.1930	GB336696	Mechanism for operating hydraulic brakes	France	Simple construction. No patent granted, fee not paid.
16.04.1931	GB346532	Improvements relating to fuel supply systems for internal combustion engines	Moorfield Works, Wolverhampton	Compression ignition engine fuel delivery relief valve
19.02.1931	GB343329	Improvements relating to liquid fuel admission valves for internal combustion engines	Moorfield Works, Wolverhampton	Hollow stem valves to avoid springs
05.03.1931	GB344209	Improvements relating to fuel injection devices for internal combustion engines	Moorfield Works, Wolverhampton	Fuel injector with means to stop foreign matter
02.04.1931	GB345788	Improvements relating to the regulation of internal combustion engines	Moorfield Works, Wolverhampton	Fuel injection valve with centrifugal regulator
12.03.1931	GB344651	Improvements relating to the controlling of internal combustion engines	Moorfield Works, Wolverhampton	c.i.e. fuel pressure control system
05.03.1931	GB344217	Improvements relating to the control of fuel inlet valves	Moorfield Works, Wolverhampton	
04.06.1931	GB349786	Improvements relating to fuel pumps	Moorfield Works, Wolverhampton	Lubricating fuel pump
14.06.1935	FR782882	Dispositif de changement de vitesses automatique à commande hydraulique	Seine	Automatic gearbox with hydraulic control
20.01.1937	FR807710	Installation perfectionnée de commande des volets d'intrados et autres dispositifs de freinage aérodynamique ...	Seine	Hydraulic controls for air brakes
17.10.1936	FR804174	Dispositif de commande hydraulique à distance		E.g. Hydraulic control of carburettors
16.02.1937	FR808810	Pompe a liquide plus particulièrement utilisable pour l'alimentation des commandes et transmissions hydrauliques		
16.02.1937	FR808810	Perfectionnements aux soupapes d'admission de combustible des moteurs à combustion interne		Eliminating side pressure on injectors

Appendix Two

Publication Date	Patent No.	Title	Louis Coatalen address	Notes
03.03.1937	FR809437	Perfectionnement aux vérins et appareils analogues		Telescopic hydraulic or pneumatic device for controlling undercarriage
05.04.1937	FR811032	Dispositif perfectionné d'alimentation des moteurs à combustion interne		Automatic controls to adapt injection to large speed variations
20.05.1937	FR812938	Pompe perfectionnée pour liquides combustibles et autres		Improved fuel injection pump to regulate recirculation pressure
04.11.1937	FR820143	Perfectionnements aux moteurs à combustion et à explosions		2 inlet, 2 exhaust, inclined valve cylinder head design for c.i.e.
28.11.1938	FR49162	Dispositif de commande hydraulique à distance		Addition to 804174 power steering
22.12.1938	FR835496	Appareil transmetteur pour commande hydraulique à distance et installation…		Hydraulic control under constant pressure at height to operate e.g. radio aerial
30.12.1938	FR835760	Installation de servo-manœuvre hydraulique à distance		Hydraulic brake application
25.07.1939	FR844436	Pompe perfectionnée		Auto-regulating pump
16.12.1940	FR859302	Mécanisme transmetteur pour commande hydraulique à distance		Reversible transmitter pump. (applied for 05.05.1939)
20.12.1941	FR51205	"		Addition to above with improved seal
05.01.1943	FR877859	Compresseur à gaz		
24.10.1950	FR967014	Accumulateur de liquide		
25.06.1952	GB674515	Improvements in liquid pressure servo braking systems for vehicles	7 rue Lesueuer, Paris	2 or more pressure cylinders, independent circuits
22.01.1952	FR998684	Installation de servo-commande et applications		Independent servo pump brakes
12.11.1952	GB682480	Improvements in liquid reservoirs		Glass or polymethacrylate
31.01.1952	FR999463	Installation hydraulique de freinage		
28.02.1952	FR1001861	Machine hydraulique, utilisable en particulier comme moteur ou compresseur		
29.05.1952	FR1009421	Installation hydraulique perfectionnée de freinage		Variable braking force reflecting load weight
21.08.1952	FR1014789	Raccord pour tuyauterie et tuyau de ce raccord		

Louis Coatalen Patents

Publication Date	Patent No.	Title	Louis Coatalen address	Notes
01.12.1953	FR1045724	Maitre-cylindre perfectionné pour installation hydraulique de commande		
23.02.1954	FR1055956	Transformateur de pression et installation hydraulique de commande en comportant application		
24.02.1954	FR1056163	Soupape perfectionnée		
25.02.1954	FR1056245	Installation perfectionnée de freinage		
19.10.1954	FR1075703	Dispositif récepteur d'installation de freinage et joint pour ce dispositif		
19.10.1954	FR1075702	Raccord pour canalisation d'installation de freinage		
22.06.1955	FR1096551	Compensateur d'usure et ses applications		Hydraulic wedges
04.10.1955	FR1101244	Frein perfectionné		
31.10.1955	FR1103115	Dispositif pour la mise sous pression de circuits hydrauliques de commande		
17.01.1956	FR1108736	Mécanisme pour la commande en particulier d'un inverseur de marche		
12.01.1956	FR1108355	Liquide perfectionné pour installations hydrauliques		Hydraulic fluid, castor oil, polypropylene glycol
29.03.1956	FR1113482	Contacteur perfectionné en particulier pour appareil lumineux de signalisation de freinage		
25.05.1956	FR1117719	Servo frein perfectionné		
25.05.1956	FR1117743	Mécanisme de servo-commande hydraulique pour direction de véhicule.		
23.10.1956	FR1125022	Installation hydraulique perfectionnée de servo-commande et contacteur électrique		Pressure failure warning light
25.06.1957	FR1139064	Piston pour cylindre de frein.		
25.06.1957	FR1139063	Piston perfectionné pour cylindre de freinage		
14.11.1957	FR1146692	Installation hydraulique de commande à distance		
31.03.1958	FR1153970	Frein hydraulique perfectionné et installation de freinage...		
27.05.1958	FR1157131	Dispositif hydraulique de commande pour boite de vitesses et similaires		

Appendix Two

Publication Date	Patent No.	Title	Louis Coatalen address	Notes
03.12.1959	FR1197842	Coupelle d'étanchéité pour piston de cylindre hydraulique		
14.12.1959	FR1199359	Maitre-cylindre pour installation hydraulique de commande et installation		
28.01.1959	FR1171575	Maitre-cylindre perfectionné pour installation hydraulique de freinage		
23.11.1959	FR1196173	Dispositif de fixation d'un réservoir d'alimentation pour maitre-cylindre d'installation hydraulique de commande		
23.03.1960	FR1212400	Réservoir en matière plastique et a goulot		
17.05.1960	FR1219362	Maitre-cylindre perfectionné pour installations hydraulique de commande de freins		
19.08.1960	FR1227387	Dispositif d'actionnement à cylindre et piston et a fluide sous pression		
19.08.1960	FR1227386	Frein à disque		
02.09.1960	FR1229040	Frein perfectionné du type a disque		
16.09.1960	FR1230553	Dispositif récepteur a cylindre et piston pour frein		
27.01.1961	FR1252161	Frein à disque perfectionné		
27.01.1961	FR1252160	Dispositif de sécurité pour installation de freinage		
28.04.1961	FR1259563	Réservoir d'alimentation en matière plastique et a goulot fileté…		
23.06.1961	FR1264626	Dispositif hydraulique de commande		
11.08.1961	FR75757	Maitre-cylindre perfectionné pour installations hydrauliques de commande de freins		
18.08.1961	FR1269737	Frein à disque perfectionné		
01.12.1961	FR76776	Dispositif d'actionnement à cylindre et piston et a fluide sous pression		
03.11.1961	FR127013	Dispositif de commande à réglage automatique pour frein à commande manuelle		
17.11.1961	FR1276052	Lave-glace		
17.11.1961	FR1276052	Installation hydraulique de commande pour véhicule…		
08.01.1962	FR77016	Frein à disque		
16.02.1962	FR1284633	Installation hydraulique de commande pour véhicule		
09.03.1962	FR1286818	Frein perfectionné à disque		
30.03.1962	FR1288940	Installation de freinage pour véhicule		
04.05.1962	FR1292340	Caissette de magasinage		
11.05.1962	FR1293166	Servo-dispositif pour installation hydraulique de freinage		
05.10.1962	FR78983	Frein à disque		
19.10.1962	FR79075	Lave-glace		

Sources

Books

Bastow, Donald, *W.O. Bentley – Engineer*, Haynes, 1978

Bentley, W.O., *An Autobiography*, Hutchinson, 1958

Bentley, W.O., *The Cars in my Life*, Hutchinson, 1961

Blight, Anthony, *Georges Roesch and the Invincible Talbot*, Grenville, 1970

Blight, Anthony, *The French Sports Car Revolution*, Foulis, 1996

Boddy, William, *The History of Brooklands Motor Course*, Grenville, 1957

Borgeson, Griffith, *The Classic Twin-Cam Engine*, Dalton Watson, 1981

Bradley, W.F., *Motor Racing Memories 1903–21*, Motor Racing Publications, 1960

Brew, Alec, *Sunbeam Aero-Engines*, Airlife, 1998

Chabot, Carole de, *Rebelle avec l'aide de Dieu*, Memorizon, 2012

Clausager, Anders, *Wolseley*, Herridge & Sons, 2016

Cliff, Norman, *My Life at the Sunbeam 1920–1935*, Ashley James, 1987

Davis, S.C.H., *My Lifetime in Motorsport*, Herridge & Sons, 2007

Demaus and Tarring, *The Humber Story*, Alan Sutton, 1989

Dowell, Bruce, *Sunbeam the Supreme Car*, Landmark, 2004

Dowell, B. and A, Richens, *Sunbeam the Brass Period*, Alan Richens Publishing, 2015

Duncan, H.O., *The World on Wheels*, Duncan, 1926

Ellis, Chris, *KLG from Cars to Concorde*, Smiths Industries, 1989

Faurès Fustel de Coulanges, Sébastien, *Fiat en Grand Prix*, ETAI, France, 2009

Favre, Eric, *Les Grands Prix Automobile de Lyon*, SEPEC, 2014

Furnival, L. (ed.), *Hobson, a Personal Story of Fifty Years*, S.D. Toon, 1953

Georgano, Nick, (ed.), *The Beaulieu Encyclopedia of the Automobile*, The Stationery Office, 2000

Harten, M. von and M. Marston, *Man of Wolverhampton*, Combe Springs Press, 1979

Hartmann, Gerard, *Clément Bayard – Pionnier Industriel*, ETAI, 2013

Heal, Anthony S., *Sunbeam Racing Cars 1910–1930*, Haynes, 1989

Hodges, David, *The French Grand Prix*, Temple Press, 1967

John, Augustus, *Chiaroscuro*, Jonathan Cape, 1952

Karslake, K. and I. Nickols, *Motoring Entente*, Cassell, 1956

Karslake, Kent, *Racing Voiturettes*, Motor Racing Publications, 1950

Kelly, Robert, *T.T. Pioneers*, The Manx Experience, 1996

Kimberley, Damien, *Coventry's Motorcar Heritage*, The History Press, 2012

Sources

Louche, Maurice, *Un Siecle de Grands Pilotes Français*, Editions Louche, 1995

Massac Buist, Hugo, *The History and Development of the Sunbeam Car 1899–1924*, Sunbeam Motor Car Company, 1924

Mathieson, T.A.S.O., *Grand Prix Racing 1906–1914*, Connaisseur Automobile AB, 1965

Nye, James, *A Long Time in Making, the History of Smiths*, Oxford University Press, 2014

Omnès, Roparz, *Les Koatanlem, Marchands, Pirates et Patriotes Bretonnes*, Editions Sked, 2000

Pomeroy, Laurence, *From Veteran to Vintage*, Temple Press, 1956

Pomeroy, Laurence, *The Evolution of the Racing Car*, William Kimber, 1966

Pomeroy, Laurence, *The Grand Prix Car*, Motor Racing Publications, 1949, revised edition 1955

Portway, Nic, *Vauxhall Cars 1903–1918*, New Wensum Publishing, 2006

Posthumus, Cyril, *Sir Henry Segrave*, Batsford, 1961

Price, Barrie, *The Lea-Francis Story*, Batsford, 1978

Ricardo, Sir Harry, *Memories & Machines*, Constable, 1968

Rochebrune, R. de and J-C Hazera, *Les Patrons sous l'Occupation*, Odile Jacob, Paris, 1995, reprinted 2013

Scott-Moncrieff, David, *Veteran & Edwardian Motor Cars*, Batsford, 1955

Segrave, Sir Henry, *The Lure of Speed*, Hutchinson, 1932

Souvestre, Pierre, *Histoire de l'Automobile*, Dunod, Paris, 1907

Spitz, Alain, *Talbot*, E.P.A., Paris, 1983

Sucher, Harry V., *The Iron Redskin*, Haynes, 1987

Taulbut, Derek, *Eagle Henry Royce's first aero engine*, RRHT, 2011

Periodicals

The Autocar
The Auto-Motor Journal
The Motor
Flight (on-line archive)
Canadian Journal of History
STD Register Journal
Motor Sport
La Vie Automobile

Unpublished Documentation

Coatalen family papers, including Olive Coatalen's scrapbooks and Marjolie Dixon Steven's draft autobiography

Anthony S. Heal Sunbeam racing car archive

STD Register Archive

Archives départementales du Finistère, Quimper

Archives Nationales, Fontainbleau

National Archives, Kew

Paul, Air Commodore, Christopher, *Louis Coatalen's Sunbeam Aero-Engines*, Fleet Air Arm Museum, 1987

Index

Note:
The following abbreviations have been used:
LC = Louis Coatalen; OC = Olive Coatalen; IC = Iris Coatalen; 'n' refers to a footnote, page numbers in italic refer to illustrations/photographs.

A

A. Darracq Co. Ltd (London) 98, 139, 140, 141, 142, 210, 211
   see also Automobiles Darracq S.A. (later Automobiles Talbot/Talbot, Suresnes, France); Talbots (formerly Clément-Talbot (London))
Accuracy Works Ltd. *96*, 97
Adams, David 252
Adams, Mr. J.W. (Sales Manager, Humber) 36, 36n16
The Admiralty 106, 107, 108, 120, 124–5, *126*
aeroengines 131, 140–1, 182, 207
   compression ignition engines 215, 217, 218, 220, 227
   early development of 87–8, 95, *109–111*, 112, *113–18*, 119, *120*, 120n13, 122–3
   Weymann's Motor Bodies (1925) Ltd. 213, 221, 221n20, 222
Aeronautical Society 120, 122–3
Air (Aero) Show (Olympia) (1913; 1929) 87, 217
Air (Aero Show) (Paris) (1920; 1936; 1938) 140, 226
Air Ministry (France) 226
'Aircraft & Motor Car Engine Design' (Aeronautical Society speech) 122–3
Aircraft Disposal Department 140
Aircraft Production Division (American Signal Corps) 131
airships 123, *124–6*, 140, 215, 217
Alcock, Sir Jack 110, 125n19
Alfa Romeo 178, 186, 251
Amalgamated Engineering Union 160
ambulances (patent) *116*
American Signal Corps (Aircraft Production Division) 131
American Society of Automobile Engineers 65, 67

Shaw, B.J. Angus 51, 52, 52n2
'Arab' engine (Sunbeam-Coatalen) 116–17, *118*
Aron, Hermann 221–2
Arrol-Johnston 69, 71, 138
Ascari, Antonio 186
Associazione Tecnica Automobile (Italy) 246
Aston, Wilfred Gordon 187
Atfield (mechanic) *179*
Attwood, S.H. 218
Aubert, Mme *71*
Aury, Jean 227
Austin, Sir Herbert 171, 199
Austin Motor Company 78, 116, 138, 212
Auto-Motor Journal 26
The Autocar 7, 15, 29, 48, 76
   articles/letters by LC 61, 160, 161, 162–3, 171, 181, 185
   Humber 31, 34, 36, 36n16
   STD Motors Ltd 147, 149, 152, 155
   Sunbeam cars 54, 74, *139*, 140
   Tourist Trophy Race (Isle of Man) 37–8, 40, 43
l'Automobile Club de France
   see Grand Prix de l'Automobile Club de France
Automobile Club of Great Britain and Ireland
   see Royal Automobile Club (RAC)
The Automobile Engineer 15, 226
Automobile Engineers, Institution of
   see Institution of Automobiles Engineers
Automobiles Darracq S.A. (later Automobiles Talbot SA/Talbot, Suresnes, France)
   200 Mile Race (Brooklands) (1925; 1926) *187*, *190*, *192*
   Bentley and *184*, 185, 188
   Grand Prix (Brooklands) (1926) 191
   Grand Prix de l'Automobile Club de France (1923) 163, 165, 168, *169*, 170, 171
   introduction of supercharging 178, 178n3, 180–1, 182, 183, *189*
   motor racing (1921) 142, 143, 145, *146*, 147, *148*, *149*, 150, 151

Index

motor racing (1922) 152, 153, 154, 155, *156*, *157*, 163
post-First World War (1918–) 193, 194, 198, 199, 212, 228, 229, 251–2
see also A. Darracq Co. Ltd (London); Talbots (formerly Clément-Talbot (London))
Automotive Industries 218
A.V. Roe and Company (AVRO) 120, 123
aviation industry 106–7
   construction of airships/aeroplanes 123, *124–5*
   development of aero-engines *109–11*, 112, *113–18*, 119, *120*
Avions Weymann-Lepère 221

B

Babs (Liberty aero-engined car) 189
badge-engineering 143
Baehr, Gustave 229
Ballot racing cars 155
Banque Française de l'Afrique (BFA) 214
Barbarou, Marius 27
Barrett, Tom (mechanic) *179*, 180, *181*, 183
Bastow, Donald 250
Bat Boat (flying boat) 123
Bate, James Bernard *73*, *91*
Bath family:
   Henry James (father of OC) 52, *52–3*, 55, *71*, *98*, 99–101, 102, *103*
   Henry (brother of OC) *173*
   Olive Mary (neé Griffiths) (mother of OC) 99, 101, *103*, *173*
Bath, Olive Mary *see* Coatalen, Olive Mary (neé Bath, second wife of LC)
Bauer Marchal 214
Bayard-Clément 26
Bayliss, N.F. 55
Bayliss, Samuel *71*, 107
Beacon House, Wolverhampton (second home of LC and OC) 107
Beardmore heavy oil engines 217, 218
Becchia, Walter 163, 199, 201
Beeston Humber 35, 36, 36n16, 39–40, 43
Belcher, Henry (Works Manager, Humber) 32
Bell, F.J. 107
Bentley Motors 175, 181, 182–3, *184–8*, 196, 198, 236
Bentley, W.O. 14, 112, 181, 215, 250

Berriman, Algernon 200
Bertarione, Vincenzo (designer) 163, 165, 168, *179*, 188, *189*, 193, 199
Berthier, Henri 231
bicycle and tricycle manufacture 29, 30–1, 43, 50, 55
Bill, Frank (mechanic/driver) 56n5, 86, 87, *144*, *179*
Billing, Noel Pemberton 123
Bird, Sir Alfred 108
Bird, Christopher A. 79, *80*, 108, 109
Birdlip Hill (Gloucestershire) 36
Black, John 49
Blackpool Race Meeting (1905) 36
Blériot, Louis 109
Blériot, Mme 79
Blight, Anthony 192, 201, 251–2
Boddy, Bill 86, 87, 250–1
Le Boeuf Couronné inn (Neuvy-le-Roi) *164*, 165
Boillot, André 143, *144*, 145, 147
Boillot, Georges 82, 83, 85
Bolles, Norman T. 220
Bordino, Pietro 165, *166*
Borgeson, Griff 15, 62
Bosch 230–1, 243
Bostock, Mr. Geoffrey (receiver) 48
Boulogne speed trials (1926) 191
Bourlier, Edmond 168, 180
Bradley, William F. *139*, 140, 145, 146, 155
Bradshaw, Granville 161
Brasier, Charles-Henri 77
Brazilian Traction 213–14
Brew, Alec 110, 111
Bridson, Ellen Mary ('Dickie') *see* Coatalen, Ellen Mary ('Dickie', fourth wife of LC)
Bridson (mechanic) 222
Brighton Speed Trials (1905) *36*
Bristol 'Phoenix' diesel engine 227
British Aircraft Constructors, Society of (1915) *see* Society of British Aircraft Constructors (1915)
British Motor Boat Club 78
British Motor Syndicate 29, 30
British Racing Motors (BRM) 199
Brittany 13, 14, 17–19, *20*, 21, *22*, 23, 55
Broadbent, Jim (mechanic) 43
Bromley House, Penn (third family home of LC and OC) 103, 129, *129*, 130, 133
Brooke Cars 78

Index

Brooklands Race Track 62, 72, *81*, *104*, 165, 171, 185
  aviation industry 110, 110n3
  LC as a driver 16, 44, 55, 56, *58*, 61, 74, *101*
  post-First World War (1919–) 143, 153
  races 80, 190–1, 196, 198
  records:
    'rating' records (Brooklands) 65
    Class F international records *186*, 187
    short-distance records 81, 84,178, *180*
    endurance records 64, 65, 66–7, 68, 69, *75*
    Toodles V 83, 95
    World Records for 1, 000 miles *75*, 80
    see also 200-mile race
Broome, Alec (mechanic) *179*, 197
Bugatti *166*, 173–4, 178, 180, 186, 199
Buist, Hugo Massac 16, 26, 68–9
Bunbury, H.W. 128, *91*
Burgess-Wise, David 200
bus production 201, 202
Butler, Edward 29

C

Cadillac 209
Caillois, Gustave *71*, *73*, 81, *82*, *83*
Caillois, Mme *73*
Calthorpes 69
Campbell, Malcolm 150, 169, 171, 189, 195, 199, 205
Capri 205, 222, 223, 228, 229, 233, *234*
Cariou, Marie Louise Clémentine (later Coatalen, paternal grandmother of LC) 19
CEPANA *see* Commission d'Examens Permanent de Projets d'Appareils Nouveaux pour l'Aéronautique
Chabot, Carole de (granddaughter of LC) 238, 240, 248
Charteris *130*
Chassagne, Jean (driver) *73*, 120, *149*
  Brooklands 80, *81*
  Coppa Florio race (Sicily) 158, *159*, *160*
  Le Mans 24 Hour Endurance Race (1925) *184*, 185
  motor boats and yachts *78*, 79, *88*
  motor racing (1913) 82, *83*, 84, *86*, *87*
  motor racing (1914) 94, *95*
  motor racing (1922) 133, 153, *154*, *155*, *156*, *157*
Château de Moreuil 81, 82, *83*
Cheesman, George 97

Le Chêne (Les Mousseaux, LC and 'Dickie's' house in France) 238, 240
Christiaens, Josef 133, 133n25
Chrysler Six motor car 228
Churchill, Winston (First Lord of the Admiralty) 106, 120
Circuit of Britain Race (1913) 110
Circuit des Routes Pavées (Lille) 181
Ciro's Club (London) 129
Citroen DS (1955) 31
Civil Order of Merit (Spain) 246
Clarke, Sir Travers 228
The Classic Twin-Cam Engine (Borgeson) 15, 62
Claudel, Henri *81*, 96, 97
Claudel, Mme 97
Clegg, Owen 166, 171, 181
  decline of STD Motors Ltd 209, 210, 212, 213, 218
  Grand Prix de l'Automobile Club de France (1921) *145*, 146, 147, 149, 150
Clément, Adolphe 26, 27, 29, 124, 142
Clement, Frank 154
Clément-Gladiator-Humber (bicycle company) 26
Clément-Talbot (London) *see* Talbots (London)
Cliff, Norman 216–17
Coatalen, Anna (second wife of Hervé Coatalen) 238, 240, *250*
Coatalen, Annie Ellen (neé Davis, first wife of LC) 42, 49
Coatalen, Betty (first wife of Hervé Coatalen) 240
Coatalen, Ellen Mary 'Dickie', (formerly Bridson, fourth wife of LC) *233*, *234*, 236, 238, 240, 248, 249, *250*
Coatalen family (France):
  Catherine (great grandmother of LC) 19
  Delphine (great-aunt of LC) 19
  Emile (uncle of LC) 20
  Ernest (uncle of LC) 20
  François Marie (father of LC) 20–1, 28, 42
  François Marie (older brother of LC) 21, 23, 24, 27n16
  Louise Marie Angélique (neé le Bris, mother of LC) 20, 21, *22*, *23*, 24, 28
  Marguerite Clémence (neé Le Sauour, second wife of
  Marguerite (great-aunt of LC) 19
  Marie Louise Clémentine (neé Cariou, grandmother of LC) 19
  Martin René (paternal grandfather of LC) 19, 20
  René (uncle of LC) 20
  Yves (great-great-grandfather of LC) 19

Index

Coatalen Heal, Annik (neé Coatalen, granddaughter of LC) 7, 14
Coatalen, Hervé Louis (second son of LC) 7, *24*, 56n6 138, *164*, *173*, *193*, *203*, 204, 234
   early childhood 90, 100, 101, *103*, 107, *129*, *130*
   marriages 238, 240, *250*
   Second War War service 238, *239*, 240
   work/relationship with LC 240, *241*, *242*, *243*, 247
Coatalen, Iris Enid Florence (neé Graham, formerly van Raalte (third wife)) *94*, *130*, 137, 138, 166
   marriage to LC *172*, 173, *174*, 203, 203n1
   relationship with Yvonne Franck 204, 234, 236, 236n3
   separation/divorce from LC 204, 234, 236
Coatalen, Jean Louis (third son of LC) 56n6 103, *203*, 231, 242, *247*
   early childhood *129*, *130*, *164*, *173*
   Second World War service 238, *239*, 240
Coatalen Jnr, Louis Hervé (first son of LC) 42, 42n1
Coatalen, Louis Hervé
   general:
     birthplace and childhood 13, 21, 23, 24
     family tree *10–11*, 17–19, *20–4*
     parents *20*, 21, *22*, *23*, 24
     military service 26, *27*, 27n16 28, 42
     naturalisation as British citizen 106–8
     personal character 7, 13, 14, 147, 149, 174, 252
     physical characteristics 13, 14, 36–7, 77, 207
     portraits *12*, *21*, *139*, *252*
     drug/alcohol addictions 204, 205, 233, 240
     nervous breakdown/Capri 205, 222, 223, 228, 233
     personal investments 213–15
     life during Second World War in France
     post-Second World War life *241–7*, 248
     death 249
   private life:
     first marriage 42
     second marriage 49, *98–105*, *128–30*
     third marriage 137, 138, *172–4*, 203–4
     fourth marriage 233–6, 248, 249, *250*
   engineering career:
     assessments of LC's contribution 249, 250–1, *252*, 253
     early career and training 24-25, 26n11, 26–7
     move to England 27, 28–9, 42
     other business interests 96–7, 128, 134–5, 136, 228, *229*, 230–2
     personal competition achievements 254–61
   public duties (1914–18) 120, 122–3, 137, 138–9, 139n4, 167
   public duties (1919–) 138, 227, 233, 245, 246
   published articles/letters 61, 88–9, 90, 160, 161, 162–3, 171, 181, 185, 218
   skills/qualities as an engineer/designer 14–16, 19, 40–1, 43, 137, 220, 226
Coatalen, Marjolie (daughter of LC) 42, 42n1, 138, 172, 203, 204, 233, *234*, 239
Coatalen, Olive Mary (neé Bath, second wife of LC) 57, 79, *82*, *84*, 88, 90, 95, 204
   family background *98–9*
   marriage/family life with LC 55, *101–3*, *104*, *105*, *128*, *129–30*, 133, 138
   motor racing *49*, *59*, 64, *65*, *66–7*, *68*, *70–3*, 75, *91*, 95
   portraits *82*, *83*, *89*, *97*, *100*, *101*, *102*, *105*, *128*
   separation/divorce from LC 173
Coatanlem, Nicolas (nephew of Yann) 18–19
Coatanlem, Yann (Jean de Coetanlem) 17–18
Colomb, Commander Horatio Walcott 42n1
Commercial Motor Exhibition (1929; 1931) 227
Commission d'Examens Permanent de Projets d'Appareils Nouveaux pour l'Aéronautique (CEPANA) 223
Commission Historique de l'Automobile Club de France 251, *252*
compression ignition engines 205, 207, 215, *216*, *217*, 218–21, *222–7*, 226n25, 233, 234
Concarneau, France (birthplace of LC) 21, *22*, *23*, 28, 42
Conelli, Count Caberto 186, *187*
Cook, Harold *81*, *84*
Cook, J. *91*, *93*
Coombes, Walter 170–1
Cooper, C.H. 32
Cooper, Gladys *162*
Coppa Florio race (Sicily) 157, *158–60*
Corre, Mr (Director, Ecole de l'Arts et Métiers School (Paris)) 77
'Cossack' Sunbeam Coatalen engine *121*, 125
Coupe de l'Auto race:
   (1911) 63, *64*, 68–9
   (1912) *70–4*, 75, 76
   (1913) 80, 81, *82–3*, 84, 85
Coupe des Nations race (1913) 79
Coupe des Voiturettes (1911; 1922; 1923) 62, 63, *157*, 168, 170
Coventry *see* Hillman-Coatalen Motor Car Co. Ltd. (later

Index

Hillman Motor Company)
Coventry Motor Museum 30
Coventry Technical Engineering Society 30
Cox & King (motor boats) 105
Cozens, L.V. *149*, 170, 184
Cresswell, Leslie C. *176*, *177*
Criterion Restaurant (London) 150
Crossman, R.F. 75
Crowden, Charles (Chas.) T. (Leamington Spa) 29, *30*
Cummings, Ivy 174
Cureton, Thomas (Managing Director, Sunbeam Motor Car Company) 95, 96, 100, 108, 127, 129, 138
    pre-First World War 50, *51*, 55, 68, *71*, 76
cycle racing 14, 27

D

Daggy, Augustus Smith 23
Daimler 29, 67, 199–200
Darracq *see* A. Darracq Co. (1905) Ltd
Darracq Motor Engineering 142
Dartevelle, Luce ('Nounou') 238, 240
Daventry 2, 000 miles trial (1907) 46
Davies, C.S.L. 18
Davis, Annie Ellen *see* Coatalen, Annie Ellen (neé Davis, first wife of LC)
Davis, Sammy C.H. (journalist/driver) 15, 153, *184*
Dawn Sunbeam motor car 201
Day, Rod 25
DBA group (Ducellier, Bendix, Air Equipment) 243
de Palma, Ralph 131
de Poorter, Edward A.H. 97
de Voogt, H.W. *235*, *236*, 248, *249*
De Vries boatyard (Holland) 236
Deansley, Edward *71*, 107
'Deep Sea Patrol' flying boat triplane *121*
Delage (company) 94, *166*, 178, 180, *185*, 186, 201
Delage family:
    Gustave 231
    Louis 77, 82, 83
    Mme 79
Delahaye 201
Demaus, A.B. 30
Despujols, Victor (boat-builder) 135, *141*
diesel engine development *see* compression ignition engines
Diesel, Rudolf 227

differentials in axles 161, *162*
directorships 16, 44, 48, 68, 73, 95
    First World War onwards (1914–) 127, 128, 138, 141, 201–2, 210, 228
Divo, Albert (mechanic/driver) *185*, 189, *190*, *191*, 199
    Grand Prix de l'Automobile Club de France (1923) 165, *166*, 168, 169, 170
    motor racing (1921) 143, *144*, *148*, *149*
    motor racing (1922) *154*, *155*, *157*
Dollfus, Maurice 221
Don, Kaye 205, 206, *207*, *208*, 209, 209n5 219
Dormann, M. Paul 214–15
du Pont, E. Paul 220, 221
Dubonnet *78*, 79
Duller, George 178, 180, *185*, 196, 198
Duncan, H.O. 26, 29, 31, 31n8
Dunlop *80*
Duray, Arthur *166*
Durnford, Pat 194
Dutoit, Paul (mechanic) *156*, *159*, *164*, *168*, *169*
Dyak engine 217, 218, 222

E

Eastmead, Frederic 51, 54
École d'Arts et Métiers (School for Skills and Trades) 24n9, 25–6
École des Ouvriers et Contremaîtres (School for Workers and Foremen) (Cluny) 24–6, 28
École Primaire Supérieure (Concarneau) 23
Edge, S.F. 64, 65
eight cylinder engines 152–3, 155, 165, 178, *189*, 192
Etablissements L. Coatalen S.A. (formerly Moteurs Coatalen) *217*, 221, 228
European Grand Prix (Monza) 169, 170
Evans, Lt Col Llewelyn 221

F

Farman, Henri 109, *110*
Faroux, Charles 14–15, 26–7, 31n8, 67, *70*, 123, *155*, 222, 227, *247*
Fédération Internationale des Sociétés d'Ingénieurs des Techniques de l'Automobile 246
Fenn, Alan R. 192
Fiat 165, *166*, 168, 169, 177, 178, 227, 251

Index

fire engines 30
First World War (1914–18) *105*
   construction of airships/aeroplanes at Sunbeam 123, *124–5*
   development of aero-engines (Sunbeam) *109–11*, 112, *113–18*, 119, *120*
   LC's naturalisation as British citizen 106–8
   LC's public duties 120, 122–3
   LC's transatlantic ties 131, *132–6*
   recognition of contribution of motor industry to war effort 138–9, 139n4
   Sunbeam's financial situation 125, 126–7
   Sunbeam's trademarks/company policies 120–3
Fisher, Lord (First Sea Lord) 120
flying start records 65, 65n11, 75, 84, *188*
The Folley, Ackleton (Bath family home) 101, *103*, *173*, 204
Fort Anne Hotel (Isle of Man) 90, *91*, 153, 154
Four Litre Sunbeams 198
four-cylinder engines 31, 31n8, *33*, 34–6, *43*, 56, 88, 192
Fowler, Langley & Wright (solicitors) 106–7, 108
Franck, Yvonne (partner of IC) 204, 204n3, 234, 236
Frank, Jessie 221
French Grand Prix
   see Grand Prix de l'Automobile Club de France
Friedrich, Ernst *166*
Frigidi-Pedibus (motor boat) 88
Fromuth, Charles (painter) 21, *22*, 23n5, 28, 42
Fuchs, Judge Emil E. 135
Fujiyama III (motor boat) *79*

G

Gadzartsde l'Automobile (graduate engineers, *École d'Arts et Métiers*) 24–6, 77, 83
Gaillon Hill Climb (Normandy) *143*, 150, 191
Garros, Roland 109
Genna, Edward 51, 54, 55
Georges Roesch and the Invincible Talbot (Blight) 251–2
Giaccone, Enrico 165, *166*
Glenn, Alley & Geer (New York) 218
Goudard, Maurice 233
Goux, Jules 82, 83, 84, 85, 94
Grahame-White, Montague 140
Grand Criterium des 21 Pieds race (1913) 79
Grand Prix, British (Brooklands) (1926) 190–1, 199
Grand Prix de l'Automobile Club de France:
   (1912) 13, 69, *71*, *73*, 75, 105
   (1913) 80, *81*, 82, *83*, *84*, 87
   (1914) *93*, 94–5
   (1921) 141, 143, 145, *146*, 147, 147n13, *148*
   (1922) *152*, *155*, 156, 163
   (1923) 163, *164–71*, 166n25, 181, 204
   (1925) *185–6*
   (1927) 199
Grand Prix de l'Europe (1924) 171, 177, 178, *179*, 180
Grand Prix de Provence (1925) 184
Grand Prix des Voiturettes (Le Mans) (1921) *148*, 149
Grand Prix d'Ouverture (Montlhéry) (1927) 199
Grand Prix, Spanish (1925; 1926) 187, 191
Graphic Trophy Hill Climb (1907) 44
Grattan (mechanic) *179*
Great Horseless Carriage Co. (Coventry) (later Motor Manufacturing Company) 29
Griffiths, Frances (grandmother of OC) *103*
Griffiths, Thomas Druslyn (grandfather of OC) 99, 99n1
Grindelwald (1912) *102*, 103
Guegen, Michel 21
Guinness, Algernon (Algy) Lee 82, 90, *91*, 92, 95, *104*, 128, 154
Guinness, Kenelm Lee ('Bill') *46*, 47, 56, *103*, 112, 128, *140*, 149, 178n3, 199
   Brooklands 200 Mile Race (1921; 1922) *149*, 156
   Coupe de Voiturettes (1922) *157*
   death of Tom Barrett 178, *179*, 180
   Grand Prix de l'Automobile Club de France (1913) *81*, 82, *83*
   Grand Prix de l'Automobile Club de France (1921) 145, *146*, 147, 147n13, *148*
   Grand Prix de l'Automobile Club de France (1922) *155*
   Grand Prix de l'Automobile Club de France (1923) *164*, 165, *166*, 169, *169*
   Grand Prix de l'Automobile Club de France (1924) 171
   Grand Prix de l'Europe (Lyon) (1924) 178, *179*, 180
   K.L.G. Sparking Plugs Works 128, 149, 230, *231–2*
   motor racing (pre-1914) 85, 88, 90, *91–4*, 95
   Penya Rhin Grand Prix (Spain) 157, *158*
   San Sebastian Grand Prix (Lasarte, Spain) (1924) 180, *181*, 183, 186
   Tourist Trophy Race (check) 154, *155*
   World Records *153*
Guy, Sydney 56n4
Guyot, Albert 67, *70*, 80, *166*

H

Haibe, Ora *144*, 145
Handley Page Type O bomber 119, *120*
Harley Bank Hill Climb (1909) *49*, *54*, 55
Harold (mechanic) *179*
Harrison, Thomas (Tommy) Herbert (chief racing mechanic) *59*, *73*, *91*, *107*, 133, *164*, *168*, *179*, *197*, 208
Harrop (mathematician) 209
Hartford Cup (1927) 199
Hartley-Smith, Jack (mechanic) 134
Hawkers *149*, 150, 161
Hawthorn, Harold Percy 230
Heal, Anthony S. 7, 56n6, 86, 87, 249, 250
Heenan & Froude 142, 210
Henry Edmonds Trophy Hill Climb (1907; 1910) 44, 57–8
Henry, Ernest 15, 62, 85, 89, 112, 155, 156, 163
Henry Farman biplane 80
Herbert, G.O. 59
High Speed Solid Injection Oil Engine (Sunbeam Coatalen) 218–19
High Speed Steel & Tool Corporation (United States) 134–5, *136*
Hillman family:
   Edith (daughter of William) 49
   Margaret (daughter of William) 49
   William 42, *43*, 44, 49, 98
'Hillman-Coatalen Brooklands racer') 44–45, *46–7*, 48, *49*
Hillman-Coatalen Motor Car Co. Ltd (later Hillman Motor Company, Coventry) 43–7, 48, 48n29
'Hilo' Ignition Ltd (Datchet) 128
Hispano Suiza V8 aeroengine 114, 116
The History and Development of the Sunbeam Car 1899–1924 (Buist) 16, 26, 68–9
H.J. Mulliner (coach-building factory, Leamington Spa) 29
Hobson, Hamilton M. 96
Holder, A.J. *88*
Home Office 108
Hornsted, L. *91*
Hôtel de France (Concarneau) 21, *22*, *23*, 28
'How Racing Influences Touring Cars' (article by LC) 171
Howarth, F. (mechanic) 40
Huggins, General 176, 184, 187, 188, 192, *196*
Hughes, Hughie 131
Humber Ltd. (formerly Humber cycle works) 29, 30, *32*–8, 39, 40, *41*, 56, 67, 77

The Humber Story (Demaus and Tarring) 30
Hunter Blair, Colin (second husband of OC) 204
Huntley Walker, Arthur 146, 149, 166, 202, 210

I

Iliff, William Marklew (Company Secretary/Director, Sunbeam) 68, *71*, 95, *96*, 107, 129, 138, 160, 171
   decline of STD Motors 184, 187, 200, 202
Imperial Airship Scheme 217
'Improvements in Starting Systems for Internal Combustion Engines' (patent) (1912) 77
Indian Motorcycle Company 205, 219–21
Indianapolis 500 Mile Race 170
   (1914) 80, 84, 86, 88, 131
   (1919) 133–4, 139, 155
   (1921) 135, *136*, 143, *144*, 145
   see also United States
Institution of Automobile Engineers 103, 138, 233, 234
International Aero Exhibition (1929) 215
Irving, Captain J.S. (Jack) (designer) 194, 196, 198, 206
   introduction of supercharging 176, 177, *179*, 184
   Racing and Experimental Department, Sunbeam 138, *144*, *151*, *152*, *168*
Irving, Miss *179*
Izod, Edwin 221

J

Jackson, J.W. 196
James & Browne 100
Janvier, Mme 79
Jellicoe, Sir John 119
Jermyn Street, London (home of LC and IC) 130, 172
John, Augustus (artist) 203–4
John Marston Ltd. 139
John, T.G. 199
Jonas Woodhead and Sons (Leeds) (vehicle spring maker) 142, 210
Junior Car Club 185

K

Karslake, Kent 52
Kay, C.B. ((General) Works Manager, Sunbeam) 90, *91*, 95, *96*, 150, *168*, 171, 184, 194, 195, 200

Index

Keen Simons and Knauth 219
Klausen Hill Climb (Switzerland) 187
K.L.G. Sparking Plugs Works 128, 149, 230, *231–2*, 238, *240–4*
Kop Hill 186
Kynochs 139

L

L-head engines 52, *57–68, 78*
Labor-Picker engine (Switzerland) 62, 62n8
Lally, Stephen 210
Laly, Robert *155*
Lambert, Charles 99
Lambert, Percy 65
Lancashire Rivington Pike Hill Climb 81
Lancaster Gate, London (LC's London flat) 174
Lanchester 116
Lavender, Edward (mechanic) *197*
Lawson, Henry J. 29, 30
Le Bris family:
   Jean François Marie (maternal grandfather of LC) 21
   Jeanne Marie Perrine (née Chéolade, maternal grandmother of LC) 21
Le Bris, Louise Marie Angélique *see* Coatalen, Louise Marie Angélique, (neé Le Bris, mother of LC)
Le Mans 24 Hour Endurance Race (1924; 1925) 175, *184–5*
Le Moigne family:
   Hervé Julian (maternal uncle/guardian of LC) *24*, 28
   Jeanne (cousin of LC) 24, *25*
   Julien (cousin of LC) 24, *25*
   Marie (cousin of LC) 24
   Marie (neé Le Bris, maternal aunt of LC) *24*
Leamington Spa
  see Crowden, Charles (Chas.) T. (Leamington Spa)
Ledu (mechanic) *179*
Leicester A.C. Hill Climb (Beacon Hill) 74–5, 95
Leo, Mr Philippe *78*
Les Fresnes, Braquemont (nr Dieppe) *71*
Levy-Beeson, Marcel *121*
Liberty engine 131, 189
Light Car Grand Prix (Boulogne) (1923) 168, *169*
Liquidation Commission (United States War Department) 210
Little-Oil-Bath chain guard 50
live-axle motor cars 52, *53–4*

Lloyd George Party Fund 138–9
Lloyds & Provincial Foreign Bank 214
Lockheed 228, *229*, 230, 233, *241*, 243
London to Brighton Emancipation Run 29
London to Coventry non-stop drive (1905) *33*, *34–5*, 36
London-Manchester-London Race (aeroplanes) (1914) 110
Long Handicap (Brooklands) 80
Lorraine Dietrich 221
Loughead, Malcolm (later Lockheed) 228–9

M

Mabley Sunbeam *168*
Macdonald, Steve (mechanic) *179*, *197*
McKenna import duty 181
Maeght *83*
Makins, A.D. 170
Manville, Sir Edward 200
'Maori II' aero-engine (Sunbeam-Coatalen) *115*, 124, 125, *126*
Maple Leaf VII (motor boat) 190
Mariage, André 222
Marie Cordelière (ship) 18
Marocchi (mechanic) *179*
Marston, Charles (son of John) 50
Marston, John (Chairman, Sunbeam) 50, 51, 68, 73, 75, 100
   First World War (1914–18) 107, 121, 126, 127
   post-First World War 137–8, 139, 211
Martin, Ernest 79
Martin, Oscar 79
Martin, Percy 200
Martinuzzi, P.F. 206, 207, *208*, 209
Masetti, Count Giulio *185*, 186, *187*
Mason, A. (Assistant Works Manager, Sunbeam) 138
'Matabele' V12 aero-engine *141*, 190, *206*
Mathieson, T.A.S.O. 94
Maudsley Motor Company 89
mechanics (racing) 43, *179*, 180, 182, 183, 186, *197*
  see also under Barrett, Tom; Bill, Frank; Dutoit, Paul; Harrison, Tommy; Moriceau, Jules; Perkins, Bill
Medinger, Emil *71*, 73
Medinger, Mme Mimi *71*, 73
Menier, Jacques 86
Mercedes Benz 94, 227
Meurisse (photographer) 67
Milan Grand Prix (Monza) (1926) 191
Ministry of Aviation (France) 241

Ministry of Defence (France) 223
Mitchell, Percy 87
Moglia, Edmond 163
Mohawk engine (Sunbeam) 119
Monaco motor boat races 78, 79, 88, 89
Monnaie de Paris (Paris Mint) 251, 252
Monnier 222
Monte Carlo Rallye (1949) 241
Montgomerie, J. (manager, Accuracy Works) 96
Montlhéry Circuit (1924; 1925) 182, 184
Moriceau, Jules (Segrave's mechanic) 168, 170, 179, 188, 190, 191
   motor racing (1921–2) 147, 148, 149, 154, 155, 159
Moteurs Coatalen (later Etablissements L. Coatalen S.A.) 217, 221, 228
The Motor 33, 52, 63, 93, 111, 147, 153, 160, 161
   article by LC 88–9, 90
   200 mph world record 196, 198, 199
motor boat engines 79, 87, 88, 105
Motor Manufacturing Company (formerly Great Horseless Carriage Co.) (Coventry) 29
Motor Mills (Coventry) 29
The Motor Organizations Ltd 211
Motor Show 44, 65, 77, 83, 194
Motor Sport 86, 56n6, 250–1
Motoring Entente 211
Movietone News 206
Munitions Levy and Excess Profits Duty 127
Murat, Princess Violette 203, 204
Murray (mechanic) 179
M.Y. Karen (LC's second yacht) 200, 204

N

Napier 35, 117
National Physical Laboratory 195
National Provincial Bank 214–15
Nautilus (single-seater car) 56, 56n5, 57, 59, 61, 89
Naval Wing (later Royal Naval Air Service (RNAS)) 106
Noney 130

O

Oaks Crescent, Wolverhampton (first home of LC and OC) 107
Oates, Alex G. 215, 216, 222

Oatlands Park Hotel, Weybridge 67, 70
Ocean Rover (ex-tramp steamer) 157, 158, 159
Officier de la Légion d'Honneur (France) 246
Ogburn, Charlton 218, 219, 220
The Orchard, Wolverhampton (third home of LC & OC) 103
Orliénas, Lyon 177
overhead valve engines 160–1

P

Packard 219, 220, 227
Panhard commercial vehicle engine 223
Panhard et Levassor 26, 29, 247
Panhard, Jean 247
Panhard, Paul 247
Paris 28, 173, 174, 233
   move to Paris with IC 137, 138, 139, 140, 141
   Second World War 237, 238, 240
Parker, F.B. 120
Parkinson, Lt Col George W. 210
Parry-Thomas, John Godfrey 189, 195, 196
Pateley Bridge Hill Climb (1912) 74
patents (Great Britain/France) 194, 201, 207, 229, 240, 243, 247, 253
   aero-engines 114, 116, 117, 118–19
   compression ignition engines 218, 219, 220, 221, 224, 227
   list of LC's patents 262–70
Pathan double-decker bus 202
Pennington, E.J. 30
Penya Rhin Grand Prix (Spain) 157, 169, 170
Percy, Harold 128
Perkins, Bill 153, 155, 159, 171, 178, 179, 180, 183, 197, 206
Perrot, Henri 67, 69, 81, 83, 229, 233, 241, 245, 246, 247
Perry, Sir Percival 210–11, 211n9
Peugeot 94, 131, 135, 155, 250–1
   dominance of 80, 82, 83, 84, 85–7, 87n24, 88, 89
Platt, Maurice 251n12
Pomeroy Jnr, Laurence 86–7, 103
Pomeroy Snr, Laurence (Chief Engineer, Vauxhall) 15, 63, 76–7, 85, 103–4, 105, 139, 168, 234
Pont Croix (Brittany) 19, 20
Pont l'Abbé (Pays Bigouden–South Finistère) (boyhood town) 14, 17, 19, 20, 21, 24
Porporato, Jean 131
Powell, Edward A. (chairman, Humber Ltd) 30–1, 32, 41
Power, J. 168

Index

Pratt & Whitney 219
Pratt, F.Gordon *88*, *89*, 105, 123
Price Waterhouse 205, 210, 212, 213
Priest, W.A. (Export Manager, Sunbeam) 138
Prix de la Méditerannée race (1913) 79
Prix de Monaco race (1913) 79
Prix du Premier Pas race (1913) 79
Provot hat factory (Chazelles) 231
Pullinger, Thomas C. (General Manager, Humber Beeston) 32, 35, 36, 40, 41, 50, 51, 52, 67, 69, *70*

Q

Q cars 152
Queen's Café, Wolverhampton 138
Queen's Dolls House (Windsor Castle) 171

R

R.34 airship 124–5
RAC *see* Royal Automobile Club (RAC)
Racing and Experimental Department, Sunbeam 55, 56, 58, 140, *168*, 170, 176, 198, 206, 215–18
Radley-England Waterplane 110
RAF 1 engine 111
Raglan Cup (1910) 57, *58*
Rawlinson, Captain Toby *91*
Red Cross Society (France) 127
Registrar of Trade Marks 121, *122*
Reid, J. 32
Reliability Trial (Royal Automobile Club (RAC)) (1902) 31
Renault 241
Resta, Dario 103, 169, *170*, 171
  Coupe de l'Auto (1912) 71, *72*, *73*, *75*
  Indianapolis 500 Mile Race (1919) 133, 135, 136
  Indianapolis 500 Mile Race (1921) 143, *144*, 145
  introduction of supercharging 177, *178*, 178n3 *179*
  killed at Brooklands (1924) *180*, 183
  motor racing (1913) 81, *82*, *83*
  motor racing (1914) 86, 87, 88, 90, *91*, 92, *94*, 95
  as Sunbeam representative in US 139
Resta, Mme Eva *71*, *83*, *91*
Resta Motors (New York) 135
Ribeyrolles, Paul 56
Richards, T.H. 64, 65, *68*
Rickenbacker, Eddie 131, 132, *133*, 144

Ridley, Jack *153*, *164*, *179*, *197*
Rigal, Mme Adda *70*, *71*, *73*
Rigal, Victor 67, *70*, *71*, *72*, *73*, 86, 90, *94*
Robertson, Andrew 192, *193*, 209–10
Robinhood Engineering Works Ltd. 128, 230
Roblin, Mr. Luc 21n4
Roe, H.V. 120
Roesch, Georges 181, 192–3, 201, 251, 251n12
Rolland Pilain *166*
Rollo, Maurice 233
Rolls, Charles 40
Rolls-Royce 40, 111n5, 184, 235, 241
Root, Marshal J. 135
Rootes Securities Ltd 210, 233
Rose, (Raymond) Hugh (designer) 36–7, 42, 56, 56n4, 56n6, 86, 202, 206, 209
Rover 41n23
Royal Aircraft Establishment (Farnborough) 226–7
Royal Automobile Club (RAC) 31, 55, *59*, 65, 72, *74*, 75, *167*, 196
Royal Flying Corps 106
Royal Naval Air Service (RNAS) (formerly Naval Wing) 106, 111, 112
Royce, Sir (Frederick) Henry 86, 86n21, 112, 228
Rudge Whitworth 46, *47*
Russia 188, 126n21
Rutter, Thornton (Daily Telegraph) *91*
Ruxton front-wheel-drive car 209

S

S. Smith & Sons (Motor Accessories) Ltd 230
St Athanase mental home (Quimper) 21
St Maurice-sur-Dargoire 94
Sainturat, Mr. 77
Salamano, Carlo *166*, 169
Salon de l'Aéronautique, Paris (1951) *243*
Saltburn Sands 95
San Sebastian Grand Prix (Lasarte, Spain) (1924) 180, *181*
Santos-Dumont, Alberto 123
Saunders, S.E. 105, 123
Sauour, Marguerite Clémence Le (second wife of Martin René (grandfather of LC)) 19
Scales, Jack (mechanic/driver) *179*, 180, 182, 183
Scotsman car 202
Scott-Moncrieff, Bunty 29

Index

Scottish Reliability Trial (1905; 1909) 34, 52, *52–3*, *54*
Searight, Major T.P. *71*, *75*, *96*, 97
Second World War 237–8, *239*, *240*
Sedwick, Charles 80
Segrave, Major (later Sir) Henry O.D. (driver) 15–16, 181, 183, 189, 191, *193*, 199
   Class F international records (Brooklands) *186*, 187
   Coppa Florio race (Sicily) *158*, *159*, *160*
   flying start kilometre at 152.3 mph World Record *188*
   Grand Prix de l'Automobile Club de France (1923) *164*, 165, *166*, *167*, *168*, *169*
   Grand Prix de l'Automobile Club de France (1924) 171
   Grand Prix de l'Automobile Club de France (1925) *185*, 186
   Grand Prix de l'Europe (1924) 178, *179*, 180
   Le Mans 24 Hour Endurance Race (1925) *185*
   Montlhéry Circuit (1924) 182
   motor racing (1921) 143, 145, 146, 147, 147n13, *148*, *149*, 150
   motor racing (1922) *154*, *155*, *156*, *157*
   Six Hour Endurance Race (Brooklands) (1927) 196, 198
   200 mph world record 194, *195*, *196*, *197*, *198*
   200 Mile Race (Brooklands) (1925; 1926) 186, 187, 190, 191
   World Land Speed Record (1926; 1929) 189, *196*, *197*, *198*, 205, 208
Semphill, Lord 249
Service Technique de l'Aéronautique (French Ministry of Defence) 223
SFFHL see La Société Francaise des Freins Hydrauliques Lockheed (SFFHL)
Shaw, B.J. Angus see Angus Shaw, B.J.
Shelsley Walsh Hill Climb 36, 55, 57, 79, 80, 108, 109, 181, 186
Short Brothers 119, 120, 123
Short Handicap (Brooklands) 80
Shorter, L.J. (Chief Draughtsman, Sunbeam) 119, 138
Shrewsbury and Talbot, Earl of 142
SIA see Société des Ingénieurs de l'Automobile (SIA)
Sicily 157, *158–60*
Sikh double-decker bus 202
'Sikh III' engine 217–18
Sikorsky Ilya Muromets V biplane (Russia) 119
Silver Bullet 205, *206–8*, 209, 219, 220
Singer *Bunny III* 65
Singer *Bunny Junior* 59

Sitges racetrack (Barcelona) *170*
six-cylinder engines 175, *176*, *177*, *178*, *181*, 192
Skegness Sands 36
Skoda 222
Slater, Dick (mechanic) *197*
Smallwood, A. Eddie 'Chips' 134–5, 136, 139
SMCC see Sunbeam Motor Car Company Ltd of Wolverhampton (SMCC)
Smith, Harold Nelson 44, 55
Smith Kirby 231
Smith (mechanic) *73*, *91*
Snake Hill (Derbyshire) 36
'Snark' motor car 54
Société des Ingénieurs de l'Automobile (SIA) 227, 233, 245, 246
Société des Transports en Commun de la Région Parisienne 222–3
La Société Francaise des Freins Hydrauliques Lockheed (SFFHL) 228, 229, 243, 245
Society of Automotive Engineers (STA) (Spain) 246
Society of British Aircraft Constructors (1915) 120
Sopwith Aviation 123
South Harting Hill Climb (1905) 36
Souvenir of Sunbeam Service 1899–1919 booklet (1919) 26, 113
Spanish Grand Prix (1925; 1926) 187, 191
speed record (public roads) 191
STA see Society of Automotive Engineers (STA) (Spain)
standing start (World Record) (1922) *153*
Staner, H. Walter (editor) *33*, *34*, *35*
Stanley, Sir Arthur 167
'The State of the Motor Industry in Coventry' (*The Autocar*) 48
STD Motors Ltd 136, 141–3, *183*
   decline of motor racing business 175, *188–194*, 200, *201*, 202
   introduction of supercharging 175, *176–84*
   LC's period of illness in Capri 228, 233
   motor racing (1921) *144–6*, 147, *148–51*
   motor racing (1922) 160–1, *162*, 163
   success against Bentley at Le Mans *184–8*
   trading difficulties 205, 209–13, 218, 252–3
   200mph world record 194, 195–9, 200
   see also Sunbeam Motor Car Company Ltd of Wolverhampton (SMCC); Talbot Darracqs (Suresnes, France); Talbots (formerly Clément-Talbot (London))

Index

Sterling Engine Co. (United States) 131, 134
Stevens, Herbert C.M. (chief designer/assistant engineer to LC) 114, 119n11, 141, 145, 161, *168*, 176, 184, 194
Stirk, Frank A. 107
Straker-Squire 61
Strothers, Isaac *73*
Sucher, Harry 219–20
Sueter, Captain Murray 106, 108
Sunbeam Despujols I (motor boat) *141*
Sunbeam hydroplane *78*
Sunbeam (Moorfield) Orchestral Society 138
Sunbeam Motor Car Company Ltd of Wolverhampton (SMCC) 14, 15–16, 31n8, 32, 41n23, 49
   ambitious programme of racing and development (1913–1914) 87, *88–96*
   boats and planes 77, *78–80*
   competition with Peugeot 84, 85, 86, *87*
   construction of airships/aeroplanes during First World War 108, 123, *124–5*
   Coupe de l'Auto/Grand Prix races (1912; 1913) 68–9, *71–4*, 75–7, 80, *81–4*
   development of aero-engines *109–11*, 112, *113–18*, 119, *120*, 140–1, 182, 190, 215
   development of motor cars (1909–10) 52–4, 55, *56*
   development of T-head/L-head models (1910–11) 52, *57–68*
   financial situation during First World War 125, 126–7
   origins 50, *51*, 98
   other business interests 96–7, 219
   post-First World War period (1914–) 137–8, *139–41*, 211
   trademarks/company policies 120–123
   transatlantic ties during First World War 131, *132–6*
   see also STD Motors Ltd
Sunbeam Racing Cars 1910–1930 (Heal) 7
Sunbeam Sports Tourer *139*
Sunbeam-Aero Engines (Brew) 110, 111
Sunbeam-Coatalen (Wolverhampton) 140, *141*, 190, 215, 217
   First World War (1914–18) *115*, 116, 119–20, 120n13, 121, *122*, 124–5, 131, 135
Sunbeamland Cycle Factory 50
superchargers 175, *176–84*, *186*, *187*, 189, 190, *191*, 198, 202
   Silver Bullet 205, *206–8*, 209, 220
Supermarine 123
Swiss Voiturette Grand Prix (1924) 178n3

T

T-head engines 52, *57–68*, 62n8
Talbot-Darracq *see* Automobiles Darracq S.A. (later Automobiles Talbot/Talbot, Suresnes, France)
Talbot (formerly Clément-Talbot (London)) 26, 139, 142, 147, 150, 151, 161, *162*, 187, *188*, 192–3, 194, 202, 210, 212
   see also A. Darracq Co. Ltd (London); Automobiles Darracq S.A. (later Automobiles Talbot/Talbot, Suresnes, France)
Tarring, J.C. 30
Taylor (mechanic) *179*
Tellier, Alphonse *78*, *121*
Tellier, Mme 79
Thisleton, Mr (Dunlop) *80*
Thomas, René (driver) *143*, *144*, 145, 147, *148*, 149, 150, 155, *166*
Thorneycroft, T. *130*
Three-Litre Super Sports Sunbeam 175, 181, 184, *185*, 187, 196, 198, 202
350 hp Sunbeam racing car 141, *143*, 152, *153*, 161, 171
Tiger Sunbeam car 206, 207
Tirrena (LC's yacht) *235*, *236*, 248
Todd, James 167, 188, *196*, 200, 201, 202, 208, 210, 211
Toodles 57, 58, 59
Toodles II 58–9, *60–1*, *62*, 65, 85, 89, 89n26
Toodles IV 64, 65, *66–7*, *68*, *69*, 74, 75, 80, 131
Toodles V 80, 83, 95, 110, 131
Toto (motor boat) *88*, 89
La Tour d'Argent (Parisian restaurant) *247*
Tour de France 171
Tourist Trophy Coventry Humber (1906) 37, *38*, 39–40, *41*
Tourist Trophy race (Isle of Man) 43, 202
   (1905) 32, 36n16, 183
   (1906) 37, *38*, 39–40, *41*
   (1908) *45*, *46*, *47*, 48
   (1914) 14, 87, 88, *90–1*, 92
   (1922) *151*, 152, *154–5*, 157
Le Travailleur 238
Tréminou fun fair (bike races) 14
Triumph 228–9
Trylona (LC's yacht) 248, *249*, *250*
twenty-day reliability trial (Russia) (1925) 188
200 mph world record 189, 194, *195–9*, 200
200 Mile Race (1922; 1923; 1924; 1925; 1926) 149, 150, *156*, 157, 168–9, 178, 178n3, 180, *187*, *190*, *191*, *192*
   see also Brooklands Race Track
Tyreless IV (motor boat) 105

Index

U

l'Union Commerciale & Industrielle of Paris 214
United Motor Industries Ltd 48
United States 131, *132–6*, 139, 205, 219–21
  see also Indianapolis 500 Mile Race

V

V8 Sunbeam aeroengines 87, *109*, *110*, *111*, 112, 113, 124
V8 Sunbeam marine engine 77, *78–9*, *80*, 113n7
V8 Sunbeam-Coatalen 'Crusader' aero-engine 124
V12 aeroengines 87, 88, 110, 112, *113*, *114–19*, *206*
van Raalte family:
  Charmian (daughter of IC) *130*, 172, *173*
  Gonda (daughter of IC) *130*, 172, 172n2 *173*
  Marcus Noel (driver and former husband of IC) *94*, 128, 129, *130*, 131, 172, *173*, 230
van Raalte, Iris *see* Coatalen, Iris Enid Florence (formerly van Raalte), third wife of LC)
Vauxhall Motors 63, 69, 76–7, 83, 103, 139
Védrine, Hubert 109
Vickers 195
Victor, Maurice 223, 224
Vicuna III and *IV* (motor boats) 79, 88
La Vie Automobile 77, 140
'Viking' Sunbeam-Coatalen engines 120, *121*
Villiers Component Co. 50
Voisin *166*
Voiturette Grand Prix (1923) 170

W

W & G Du Cros Ltd (Acton) 142
W. White & Sons (Cowes) *140*
Wagner, Louis 186, 198, 199
War Department (Liquidation Commission) (United States) 210
War Office 108, 110, 114, 126
Waverley House, Wolverhampton (fourth home of LC and OC) 86, *87*, 103, 129, 173
Weymann, Charles 221, 221n20, 222
Weymann's Motor Bodies (1925) Ltd. 213, 221, 221n20, 222
White, John Samuel 123
Wilder, Mr Laurence 218–19, 220
Wilding, Henry 206, 207–8, 209, 209n6
Wilks, Spencer 49
William Morris Company 212
Williams 198, 199
Willys Overland 117
Wishart, T.D. (designer, Crossley) 161
W.O. Bentley – Engineer (Bastow) 250
Wolseley-Siddely 62, *63*, 67, 212
Wolverhampton
  see Sunbeam Motor Car Company Ltd of Wolverhampton (SMCC); Sunbeam-Coatalen (Wolverhampton)
Wolverhampton A.C. Hill Climb (1904) 100
Wolverhampton District Automobile Club 55, *59*
World Land Speed Record 13, 56, 137, 157, 171, 189, 195, *197*, *198*, 202
  Silver Bullet 205, *206–8*, 209, 219
World Wars *see* First World War; Second World War
Wright, C.N. 201
Wright, R.M. (agent, Humber) 36, 38, 38n19
Wright, Warwick (sales agent, STD Motors) 171
Wyer, John 193, 250–1

Y

Youki (motor boat) 88
Younger, Mr Justice 121

Z

Zaneta (steam-powered yacht) *140*, 173
Zborowski, Louis 170, 183
Zeppelin L33 airship 124
Zone Industrielle du Pont d'Arcole, Beauvais 245, 245n8
Zuccarelli, Paul 85

UNICORN

First published by Unicorn
an imprint of Unicorn Publishing Group LLP, 2020
5 Newburgh Street
London W1F 7RG
www.unicornpublishing.org

All rights reserved. No part of the contents of this book may be reproduced, stored in or introduced into a retrieval system, or transmitted, in any form or by any means (electronic, mechanical, photocopying, recording or otherwise), without the prior written permission of the copyright holder and the above publisher of this book. Every effort has been made to trace copyright holders and to obtain their permission for the use of copyright material. The publisher apologises for any errors or omissions and would be grateful if notified of any corrections that should be incorporated in future reprints or editions of this book.

Text © Oliver Heal, 2020

10 9 8 7 6 5 4 3 2 1

ISBN 978-1-912690-69-5

Designed by Felicity Price-Smith
Printed in the EU on behalf of Jellyfish Solutions